THE SINGULARITY PARADOX

ANDERS INDSET FLORIAN NEUKART

THE SINGULARITY PARADOX

BRIDGING THE GAP BETWEEN HUMANITY AND AI

WILEY

Published by John Wiley & Sons, Inc., Hoboken, New Jersey.
Published simultaneously in Canada.

For general information on our other products and services or for technical support, please contact our Customer Care Department within the United States at (800) 762-2974, outside the United States at (317) 572-3993 or fax (317) 572-4002.

Wiley also publishes its books in a variety of electronic formats. Some content that appears in print may not be available in electronic formats. For more information about Wiley products, visit our web site at www.wiley.com.

Library of Congress Cataloging-in-Publication Data is Available:

ISBN 9781394309641 (Cloth)
ISBN 9781394309658 (ePub)
ISBN 9781394309665 (ePDF)

Cover Design: Paul McCarthy
Cover Art: © Getty Images | Eugene Mymrin

SKY10099583_030725

"A thinker without a paradox is like a lover without passion: a paltry mediocrity."

—Søren Kierkegaard, Philosophiske Smuler (1884)

Contents

Preface: The Final Narcissistic Injury of Humankind

"Subdue the Earth!" With these resounding words, the divine command of the Old Testament spurred humanity's entrance into the world. For countless generations, people held unwaveringly to the notion that the Earth must be the center of the cosmos. Surely our loving Creator would place humanity—his most cherished creation—on a pivotal planet rather than relegating us to a dim corner of the universe. Enter Nicolaus Copernicus, the disruptor of Earth's centrality, shattering the illusions of the faithful. While this cosmic demotion was a bitter pill, it laid the groundwork for groundbreaking scientific discoveries and revolutionary technologies, from Einstein's general relativity to quantum theory, space exploration, and the James Webb Space Telescope. Our "cosmological injury" initiated a series of painful revelations that Sigmund Freud called the "narcissistic injuries" of humankind.

Fast-forward three centuries, and Charles Darwin unveiled another unsettling truth: We were not fashioned in God's image but are the products of biological evolution. This "biological injury" stripped humanity of its God-given status, positioning us firmly within the animal kingdom. Our collective ego suffered another blow, but this made way for progress and new knowledge, such as genetic engineering and decoding the human genome.

In the twentieth century, humanity endured a "psychological injury," learning that unconscious drives and emotions governed us far more than rational thought, moral imperatives, or conscious willpower. This third bruise to our self-esteem led to remarkable advancements, particularly in medicine and psychology.

This series of painful realizations, which have shaken human certainties to their core, now enters a final chapter in the twenty-first century. Humankind confronts its ultimate narcissistic injury—the colossal violation of our self-esteem by the mistaken belief that we can create a posthuman hyper-technology that can outperform us in every way: the Ubermensch. In the Nietzschean sense, this isn't a human endeavor but a form of self-deception, a "mind extension" whose underlying "spirituality" is neither understood nor explainable—just a disparate species that could render its progenitors *Homo obsoletus.*

In this new era, Man—the Mensch—has now taken evolution into his own hands, seeking to transcend the limitations of our biological existence. Through a technological singularity, humanity aims for nothing less than bliss, divinity, and immortality born out of the machine: *Deus ex Machina.* The created now become the creator of their creation, forging a new reality where biological constraints no longer define our potential.

This bold endeavor challenges the very nature of human existence. We journey to reshape our existence and redefine our place in the cosmos.

While Freud's three injuries inadvertently spurred progress, humanity now actively seeks advancement. We have come to believe that experimentation and progress are fated companions, so we now take human creation into our own hands. But this time, humanity may fail to benefit from the revelation. This narcissistic drive for technological progress may, in fact, doom humankind.

As we find ourselves teetering on the edge of an unprecedented frontier, our challenge lies in our relentless striving toward a dynamic equilibrium between our unyielding drive for progress and the essence of our humanity. By examining the far-reaching consequences of burgeoning technologies, fostering cross-disciplinary dialogue, and placing emphasis on the philosophical and ethical facets of our existence, we can ensure that our pursuit of progress elevates and enriches the human experience rather than dilutes or displaces it.

This tectonic shift has the potential to unfold in at least one of two ways: either through the emergence of a digital superintelligence or by humanity wholly merging with technology, intertwining every visible and invisible aspect of our existence with a digital interface. Our yearning for a Deus ex Machina—the coveted embodiment of the divine and creation itself—is seemingly within our grasp, as thinkers like Ray Kurzweil suggest, with the dawn of a technological singularity looming on the horizon.

What has been seen as fiction will soon be a fact. If the Mensch is to be kept as an entity beyond the mere explicable physical universe, now is the time to reflect on potential futures and scenarios of how a technological singularity might unfold. Suppose the Mensch now takes evolution into its own hands by hacking biology. In that case, if we are now to form and create conscious entities moving from the divine act of creationism to the act of creationism—where we gain the capability to create and shape any given future—the absolute paradox arises: If everything is credible, what will happen with the Mensch? If we create conscious entities, what remains to be referred to as human?

As we close the gap between humanity and artificial general intelligence, we are left with one interesting paradox: If we build everything and design entities that are identical, how do we ensure there is something left in the unknown for the Mensch to ponder? How do we ensure the potential of infinite progress, and how can we build and shape yet still leave that very thing that is not thinkable in the unknown?

The final narcissistic injury, however profound or wretched, should never be fully realized, as we—the Mensch—would cease to exist. This could manifest literally (as *Homo sapiens* is supplanted by its creation) or, even more tragically, when the species loses the capacity to shape its destiny. In Greek mythology, the vain youth Narcissus is captivated by his reflection in the spring. In the final narcissistic injury, the reflection would remain, but no self would be left to recognize it—the loss of perception's perception.

For many years, we have been interested in how a philosophical zombie apocalypse can be avoided and what the role of the Mensch could hold. This is our journey, the dance at the outskirts of mind and matter, where philosophy and physics meet to change, and maybe this is where we will find life: in the paradox, in this wonderful dance that makes life livable for the Mensch.

Each contradiction is a paradox because they are connected, and that is where you find life. That is where life is, in the paradox. The dance of it? Philosophy and physics.

Introduction

To this day, we humans live with two irrefutable truths: Everything is in constant change, and we all will die. The first one, *panta rhei*—"everything flows," as the Greek philosopher Heraclitus described it—explains that we cannot step into the same river twice, "for it is not the same river, and we are not the same man"; it is a realization with which we humans can cope. The second statement, on the other hand, that we all will die, is a little trickier. It triggers fear because it is a step into the unknown. Is life, then, over, or does the soul experience a transition into another being? Existential angst arises from pondering the nature of human existence, the meaning of life, and the inevitability of death.

The human being, or Mensch (in German), as we will refer to it in our book, possesses what seems to be a unique ability to be consciously aware of its mortality. This awareness can lead to existential angst as we grapple with the fact that our lives are finite and that we will eventually cease to exist. Our fear of death can drive us to seek meaning and purpose in our lives, and existential angst arises when the Mensch struggles to find a sense of meaning or purpose, leading to feelings of emptiness and despair. It is also the representation of the ultimate unknown that causes fear, because no one can honestly know what happens after we die. We may feel overwhelmed by the uncertainty surrounding death and the afterlife. Death is an inevitable part of life, and the fact that it is beyond our control can lead to feelings of helplessness and anxiety. Existential angst can stem from this lack of control as people grapple with the idea that their existence is ultimately subject to forces beyond their influence.

What is death? There are no absolute answers. Religion and research offer many building blocks, usually with a comforting message: the path to transition can be painful, but dying itself is peaceful, even for the companions. Just as each life differs, so too does the dying process. Biologically, each dying process can be defined, but only in terms of the death of individual cells or organs. Even during embryonic development, programmed cell death occurs repeatedly when excess cells that are no longer needed during development kill themselves. Later, T lymphocytes—white blood cells—eliminate cells that are harmful to the organism. Even whole organs, such as limbs or the spleen, can die while the person continues to live. Surprisingly, little research has been done on what causes an organism to die. Cardiovascular failure, often cited as a reason, is usually the result of other circumstances. Is it the soul itself, whose life energy slowly halts according to a plan, as many spiritual wisdom teachings claim? Transplant medicine is currently raising the question of whether human death can be defined as brain death. Only total brain death has been established as a necessary condition for organ harvesting.

What about life after death, which religions often promote? No one knows the truth; therefore, there are many interpretations of what life after death might look like. Christian religions provide surprisingly little accurate information. Rebirth is ruled out, but from an all-death theology to a personal judgment that decides the soul's subsequent whereabouts in heaven or hell, everything seems possible and nothing concrete. Most Christian theologians today assume that the whole person (with body and soul) dies at the moment of death but immediately awakens to a new life in the afterlife. Accordingly, the individual resurrection would not occur on the Last Day but immediately after each person's death. For some philosophers, separating the soul from matter is not possible.

Perhaps most difficult to understand is the view of many believers that after death, there should be a trial, a retrospection, or a negotiation in which the future whereabouts of the soul are determined. Imagine that a person has finally overcome the agony of death and is either judged immediately, leaving no further opportunity to improve on the mistakes of one's life, or the personality is preserved in an intermediate state indefinitely until the Day of Judgment. This view, though fortunately no longer so strongly taught today, was for a long time considered binding on believers, probably also to keep them from loving each other in life. Religion is always a mirror of its time.

The ancient wisdom texts should also not go unmentioned. The *Bardo Thodol,* or the *Tibetan Book of the Dead*, describes precisely and meticulously the heavenly worlds into which the soul will enter, places supposedly seen by Tibetan monks who could make such a journey while in a superconscious state through deep meditation. In many spiritual and metaphysical contexts, the superconscious state is described as a level of consciousness beyond humans' everyday waking consciousness, transcending the individual mind's limitations. It is often associated with states of enlightenment, spiritual insight, or deep intuitive understanding. It's seen as a state where one can access universal truths or wisdom, often linked with the concept of collective consciousness. Perhaps the ancient Egyptian texts describing similar afterworlds, including the so-called tunnel, were written by people with near-death experiences. We can learn a lot from religions and ancient texts, especially about the value of life.

In this book, we will follow a different path and take immortality into our own hands. Even without a scientific understanding and agreed-upon definition of consciousness, let alone a superconscious state, science is helping us more and more to understand the universe and unravel the mysteries of life, and we might soon reach a point where we can decide for ourselves if and when we want to go. In this book, we will explore the journey toward technological singularity and how perhaps consciously building singularity and creating what we will refer to as *artificial human intelligence* (AHI) might be humanity's (only) path to follow. We will look at the arising paradoxes we are confronted with when building AHIs—identical to the Mensch—yet keeping the concept of "The Mensch" unsolved.

The concept of singularity, also known as technological singularity, refers to the hypothetical future in which technological progress will become so rapid that it will fundamentally change the nature of human civilization and the world as we know it. Some people think that the singularity could be triggered by the development of artificial intelligence (AI) that surpasses human intelligence, leading to a scenario in which machines become capable of independently creating new technologies and advancing exponentially.

There is ongoing debate about whether the singularity is likely to occur and, if so, when it might happen. Some experts think the singularity is not near and may never happen, while others think it is imminent and could occur within the next few decades. There are several reasons why the singularity may not be near. One reason is that it is difficult to predict the future course of technological development and its progress rate. It is possible that technological progress may not proceed as rapidly as some people expect or that unforeseen obstacles may arise that slow down or prevent the development of the technologies thought necessary for the singularity to occur.

Another reason is that significant technical challenges need to be overcome before AI can reach the level of intelligence required to trigger the singularity. For example, AI systems still struggle with tasks that are simple for humans, such as natural language processing and understanding context. While progress is being made in these areas, it is unclear how long it will take for AI to reach the level of intelligence required to trigger the singularity.

Ethical and social considerations also need to be taken into account when considering the potential impact of the singularity. Some people are concerned about its potential negative consequences, such as the displacement of human workers by intelligent machines or the potential for AI to be used for malicious purposes.

Creating machines with conscious experiences has been one of the most difficult challenges in AI. Since the emergence of AI as a research field, researchers have been exploring ways to create machines that mimic humanlike consciousness. However, the challenge is further compounded by the numerous philosophical definitions of consciousness. These definitions range from the most superficial interpretations, such as the awareness of one's surroundings, to more complex interpretations, such as the presence of self-awareness or subjective experiences.

One of the most common starting points for understanding consciousness is examining qualia or the subjective experiences associated with sensory perceptions, such as the redness of a rose or the sweetness of a piece of chocolate. However, it is still not completely clear how the brain creates conscious content. This has led researchers to focus on understanding the neural mechanisms underlying conscious experience. To develop machines that can exhibit conscious experiences, defining a starting point for understanding consciousness is necessary. This is because we must first understand how consciousness works before creating machines that can have conscious experiences. Only then can we create hardware and software capable of reproducing the conditions required for creating conscious experiences.

This involves designing algorithms that mimic the neural mechanisms underlying conscious experience and developing machine learning models to learn from and adapt to their environment.

Over the years, scientists and philosophers from various fields have attempted to understand the creation of conscious experiences. Some have focused on the biological and cognitive mechanisms underlying conscious experience, whereas others have explored the philosophical implications of artificial consciousness. From what we can see today, some attempts have already been crowned with success, but there is still a long way to go.

When we use the term *consciousness* in this book, we refer to this as phenomenal consciousness—qualia (e.g., the subjective experience of what it feels like to be something). We are aware of multiple differences and understandings of consciousness, and the problem is that very often, there is no clear common understanding as a basis for discussions within various fields and understandings of what is meant when we talk about consciousness.

Current AI's strength lies in its ability to outperform humans in tasks that machines have historically excelled at and in its ability to reproduce aspects of humanlike consciousness. The ability to create machines that mimic humanlike consciousness is a significant milestone in the development of AI. It opens up new avenues for research and innovation, potentially revolutionizing fields ranging from healthcare to entertainment. However, there are ethical implications, such as the risks of creating machines that can think and feel like humans. Creating machines with consciousness has been a long-standing challenge in AI, and the idea becomes even more intriguing when considering that quantum physical phenomena could play a role in creating conscious experiences in living organisms. Such effects may occur on the level of cells, such as the tubulin dimers found within the nerve and other eukaryotic cells' cytoskeletons. If this is the case, it could significantly increase the number of operations a brain can accomplish per second. Although it is still uncertain whether quantum physical effects are directly linked to the creation of consciousness, it is useful to consider this possibility when designing artificial entities meant to experience conscious content.

For millennia, consciousness has been one of the most complex and intriguing topics humans have tried to understand. The term helps us discuss the different processes and states associated with the human brain, even though we cannot fully explain what it is or how it works.

Consciousness enables us to easily include as-yet-unknown processes and (changes of) states associated with the human brain in our everyday language, not only without being able to describe what accounts for a conscious experience on neuronal or (sub-) atomic layers, but also without being able to explain what consciousness is on a more abstract layer.

Consciousness has always been a challenging topic for scientists and philosophers alike. The complexity of the subject has meant that despite centuries of inquiry and study, we still need a comprehensive understanding of what consciousness is, how it emerges, and how it can be quantified or measured. Many bright minds from different fields have been working to uncover the mysteries of consciousness, and there are many theories about what it is and how it works. Most of these theories share a common feature: including a feature set associated with the perception of consciousness. However, these theories often oversimplify the complex processes involved in consciousness, such as self-reflection, deliberative thinking, and subjective experiences.

To truly understand consciousness, we need more complex theories that consider these different factors. While there is no single theory that can fully explain consciousness, researchers have proposed several approaches. One approach involves studying the brain and how it processes information to create conscious experiences, and another involves studying individuals' subjective experiences to gain insights into the nature of consciousness.

Still, even without fully demystifying consciousness, there is a clear path forward in developing artificial entities with conscious perception and taking evolution into our own hands. We call them AHIs. The concept heralds an exciting and revolutionary change in the trajectory of human evolution. By integrating the complexities of the human brain with the capabilities of synthetic components, AHIs offer the prospect of preserving and enhancing human consciousness, allowing us to transcend our biological limitations and achieve a new form of existence.

Unlike artificial general intelligence (AGI), which involves the challenge of creating consciousness from scratch in machines, AHIs start as humans and thus already possess consciousness. This intrinsic consciousness is gradually enhanced and preserved through the integration of artificial components. The development of AHIs would mark a significant evolutionary step, enabling the preservation and amplification of our conscious experience while transforming us into intelligent entities that extend beyond the constraints of our biological heritage.

The creation of consciousness in AGI raises many new questions, particularly about the nature of immortality and identity. For example, if machines can produce conscious content and theoretically live forever, some may ask whether humans could benefit from transferring their minds, dreams, and desires into these artificial vessels. While this idea may not be immediately obvious, it opens intriguing possibilities for exploration and presents ethical considerations that must be carefully weighed.

As we contemplate the future possibilities of progress, it is essential to recognize the potential for humanity to face existential risks before reaching technological singularity. There could even come a time when humanity collectively decides to halt progress due to a deep understanding or loss of interest. Nevertheless, we can move toward a singularity that aligns with human values and aspirations by creating technology based on a profound understanding of consciousness and its preservation.

The modification of our DNA and the gradual replacement of our biological bodies with artificial components, merging with AI and AIs based on organic substrates (such as the brain), will play a crucial role. In particular, despite its potential, AI poses many risks. For as long as the idea of AI has existed, there has also been the fear of it—the fear that humanity could be surpassed and then wiped out by something it created. Scientists and technologists warn that the development of AGI could end the human species. We will consider AI as both an ally and an adversary to humankind. Following this path, we will reach the technical singularity—when AI becomes so advanced that it ultimately reaches, surpasses, and merges with human intelligence.

This hypothesis is that AGI cannot be approached solely from an intelligence perspective but should also be created taking a complete understanding of biology. Our starting point is, therefore, finding ways to change and overcome humans' biological limits through technology, science, and philosophy.

With the singularity paradox, we will expand the definition of life and our understanding of the human being—the Mensch—and introduce what we can call "undead." We will argue that immortality is not a question of faith but of knowledge, which science will answer, and is a hurdle we will overcome with the help of technology. However, by doing so, the human consciousness might be at risk. The state of being "undead," or the very concept of livelihood—being alive and experiencing it with all its facets—is an idea we will explore as humanity faces a new risk that will challenge the concept of existentialism, namely that of a "zombie-apocalypse," the loss of the human conscious experience or the rise of artificially created consciousness built on a complete understanding of biology. By evolving humans into AHIs through the gradual replacement of biological brain components with synthetic ones, we enter a new era where the boundary between biology and technology blurs, raising profound philosophical questions and paradoxes.

With *The Singularity Paradox*, we aim to explore the concept of technological singularity and the evolution of humanity into AHIs. This process is not about creating artificial consciousness from scratch but about enhancing and transforming human consciousness through the integration of advanced technology with biological substrates. As we gradually replace the brain's biological components, the human body may eventually become irrelevant, leading to beings that retain human consciousness and identity but exist in a new, synthetic form.

If these evolved entities are indistinguishable from human beings in terms of consciousness and experience, they could be considered "human" in a cognitive and experiential sense, even though their origins differ from traditional biological humans. This challenges our conventional understanding of what it means to be a "Mensch" and introduces new paradoxes surrounding the nature of consciousness, identity, and continuity. The transformation into an AHI, while maintaining a foundation in biological substrates, will force us to reconsider what it truly means to be human and how we define life and existence in an era where technology and biology are inextricably linked. For example, it could be theoretically paused and restarted without any loss of consciousness. Does this artificial entity experience life and existence like a biological human? Also, if the created becomes created, the concept of divine creationism is challenged by humane creationism, which we call the Creator-Creation Paradox.

This paradox becomes particularly salient if humans create an AI entity from biological substrates that are indistinguishable from a human, because this represents the paradox of a creator (humans) crafting a creation (AI) that mirrors the creator so closely that the boundaries blur between the creator and creation, which will give birth to pronounced ethical and moral paradoxes: If the entity is biologically based and indistinguishable from a human, does it have human rights? Should it be treated like a human? Or, since it's artificially created, can it be owned or controlled?

These are challenging moral and ethical dilemmas already heavily debated in the philosophical community. The singularity paradox still stands with our chosen approach even when the artificial entity is developed based on biological substrates. This is often referred to—without a biologically based AI—as the paradox created when AI can improve its intelligence faster than humans. It could create an intelligence gap that humans may never bridge, even though it is seemingly "human" in structure and cognition. Creating an AHI indistinguishable from a human being involves navigating through layers of paradoxes, a journey this book seeks to explore in depth. The intricate interplay between biological authenticity, evolving consciousness, and societal implications forms the core of our philosophical-scientific exploration of these contradictions and paradoxes. In the midst of the technological tsunami we are currently witnessing, a new form of existentialism is emerging. We refer to this as *vita-existentialism*—a concept where the challenges posed by advanced technology, particularly in AGI and AHI, redefine the very notions of life and existence.

Our endeavor centers on the gradual augmentation and eventual replacement of the human brain, a process that leads to the evolution of AHI. This transformation is a profound continuation of self, where identity and consciousness extend beyond their biological origins, integrating with advanced technologies that redefine our understanding of existence. The journey toward AHI is far more than a technological pursuit; it is an existential evolution that challenges us to reconsider the very essence of what it means to be human. As we explore this transformation, we invite deeper discussions around the philosophical, ethical, and societal implications accompanying this revolutionary integration of humanity and technology.

1 Artificial Human Intelligence: Beyond the Human Mind

In the rapidly advancing world of artificial intelligence, we stand at a crossroads where two paths diverge. One path, widely discussed and pursued, is the creation of artificial general intelligence (AGI)—a synthetic construct that seeks to replicate or even surpass human cognitive abilities through computational prowess alone. The other path, less conventional but profoundly transformative, is the evolution of artificial human intelligence (AHI), which does not merely simulate human intelligence but represents a transformative step in human evolution itself.

AGI is the quest to build machines that can think, learn, and solve problems across a vast array of domains, mimicking the flexibility and depth of human thought. These systems, born entirely of code and silicon, are designed to process data, recognize patterns, and execute tasks with a level of precision and speed that humans cannot match. The promise of AGI is the creation of an entity that can understand and apply knowledge across all fields, potentially achieving a superintelligence that transcends our own. However, this purely artificial genesis also carries inherent risks; chief among them is the challenge of ensuring that such intelligence remains aligned with human values and does not spiral out of control.

In stark contrast, AHIs are not simply another form of AI—they are an evolved extension of humanity itself. The journey of an AHI begins within the human brain, using cutting-edge technologies like quantum nanobots, neural prosthetics, and advanced biomaterials to gradually replace biological neural structures with synthetic ones. This process is not about creating intelligence from scratch but rather enhancing and ultimately transforming human consciousness while preserving the continuity of the individual's identity, memories, and values. Unlike traditional machines, an AHI is not merely an advanced computational system; it is a conscious being that can perceive, feel, and have subjective experiences akin to those of a human. To realize an AHI, the artificial system must be capable of processing information with a level of complexity sufficient to generate subjective awareness. This involves the integration of multiple layers of processing, including sensory input, memory, attention, and decision-making.

Given the uncertainty surrounding the exact neural mechanisms that give rise to consciousness, the gradual replacement of biological components with artificial ones must be approached with extreme caution. It is imperative to maintain the intricate balance within the brain that enables the emergence of consciousness, ensuring that each step in the transformation preserves the individual's subjective awareness and identity. This careful approach is essential because the AHI, while vastly enhancing cognitive and physical capabilities, must retain the same subjective experiences and personal identity as the original human being.

The distinction between AGI and AHI is more than a difference in approach—it reflects two fundamentally different visions of the future. As a purely artificial construct, AGI raises profound ethical and philosophical questions about autonomy, control, and the nature of intelligence itself. What happens when a machine, created by humans, begins to think and act with a will of its own? Can we ensure that its goals remain aligned with ours, or do we risk creating a new form of intelligence that may one day view humanity as an obstacle rather than a partner?

AHIs, on the other hand, offer a different promise. By evolving directly from human beings, they present a vision of the future where advanced intelligence and human values are not in opposition but are deeply intertwined. The AHI is not a machine imitating humanity—it is humanity, evolved and enhanced. This path seeks not just to extend the boundaries of intelligence but to preserve and elevate the essence of what it means to be a Mensch, ensuring that our ethical and emotional core remains intact even as our cognitive capabilities expand beyond our wildest dreams.

Consider the metaphor of a ship that is gradually repaired until none of its original parts remain—a concept often referred to as the Ship of Theseus. Does the ship remain the same ship, or is it something entirely new? Similarly, as a human brain is progressively augmented with synthetic components, the individual remains the same person, but with vastly enhanced capabilities. However, the continuity of consciousness and identity is maintained, unlike in AGI, where an entirely new entity is created. In the case of AHIs, this continuity allows the individual to retain their sense of self and subjective experiences, ensuring that the transformation enhances rather than disrupts their existence. In the following chapters, we will explore the theoretical foundations, developmental pathways, and profound implications of both AGI and AHI. We will describe some of the cutting-edge technologies that make these advancements possible, from quantum computing to neuroscience, and consider the philosophical and ethical questions they raise. By examining the contrasts between AGI's synthetic origins and AHI's evolutionary approach, we aim to shed light on the potential benefits, risks, and the very different futures these two paths represent.

Theoretical Foundations of AGI and AHI

The pursuit of AGI and AHI are two distinct paths in the quest for advanced intelligence. AGI is rooted in the ambition to create machines that can perform any intellectual task a human can, with the ability to generalize knowledge across different domains. The conceptual foundation of AGI is based on achieving human-level intelligence through purely artificial means, relying heavily on machine learning, deep learning, and neural networks. These technologies enable systems to learn from vast amounts of data, recognize patterns, and make decisions autonomously. Unlike narrow AI, which is designed for specific tasks, AGI aims to replicate the broad cognitive abilities of the human brain.

Key components of AGI include the following:

- **Learning Algorithms:** This encompasses various techniques, including supervised learning, unsupervised learning, reinforcement learning, self-supervised learning, and neural networks (such as deep neural networks, convolutional neural networks [CNNs], recurrent neural networks [RNNs], and transformers).

- **Cognitive Architectures:** Frameworks such as SOAR (Strengths, Opportunities, Aspirations, and Results), ACT-R (Adaptive Character of Thought—Rational), and OpenCog that integrate various cognitive functions like memory, perception, and decision-making.
- **Natural Language Processing (NLP):** Techniques for understanding and generating human language, including transformers and language models like GPT (Generative Pre-trained Transformer).
- **Knowledge Representation:** Methods for storing and utilizing knowledge in a form that AGI can process, including semantic networks, ontologies, and knowledge graphs.
- **Reasoning and Logic Systems:** Mechanisms for formal reasoning, including symbolic AI, logic-based systems, and probabilistic reasoning.
- **Learning Transfer and Meta-Learning:** Techniques that allow AGI to transfer knowledge between domains and learn how to learn.
- **Perception Systems:** Components for processing sensory data, such as vision systems, audio processing, and sensor fusion.
- **Planning and Decision-Making:** Algorithms for generating and evaluating plans and making decisions based on objectives and constraints.
- **Self-Improvement Mechanisms:** Processes that enable the AGI to refine and improve its own algorithms and architectures over time.
- **Ethics and Value Alignment:** Systems designed to ensure that the AGI's actions are aligned with human values and ethical principles.
- **Scalability and Computational Infrastructure:** High-performance computing resources, including Graphics Processing Units (GPUs), Tensor Processing Units (TPUs), and potential future quantum computing systems, to support the processing needs of AGI.
- **Embodied AI and Robotics:** Integrating AGI with physical systems, allowing it to interact with and manipulate the physical world through robotics and autonomous systems.
- **Human-AI Interaction Interfaces:** Methods for seamless interaction between AGI and humans, including user interfaces, conversational agents, and augmented reality systems.

- **Memory Systems:** Long-term and working memory systems that store and retrieve information efficiently, enabling AGI to recall past experiences and use them in decision-making processes.
- **Social and Emotional Intelligence:** Systems that enable AGI to understand, interpret, and respond to human emotions and social cues, allowing for more natural and empathetic interactions.
- **Autonomous Learning and Exploration:** Mechanisms for AGI to autonomously explore new environments, acquire knowledge, and learn from its surroundings without human supervision.
- **Causality and Counterfactual Reasoning:** The ability to understand cause-and-effect relationships and reason about hypothetical scenarios is critical for robust decision-making and problem-solving.
- **Multi-Agent Systems:** Techniques for AGI to interact and collaborate with other AI agents or humans, including coordination, negotiation, and competition in multi-agent environments.
- **Security and Robustness:** Ensuring that AGI systems are secure, resistant to adversarial attacks, and robust against errors or unexpected inputs.
- **Lifelong Learning:** The capability of AGI to continually learn and adapt over its lifetime, integrating new knowledge without forgetting previous experiences (overcoming catastrophic forgetting).

AGI aims to create systems that are not only intelligent but also capable of self-improvement. As these systems learn and evolve, they have the potential to surpass human intelligence, leading to the concept of superintelligence. However, this potential also raises significant concerns about control, alignment with human values, and the risks associated with autonomous decision-making. Imagine a scenario where an AGI, initially programmed to optimize a company's logistics, autonomously decides that reducing human involvement is the most efficient path, leading to unintended consequences. Such examples highlight the ethical and existential risks of AGI, where a misalignment of goals could lead to outcomes that conflict with human well-being.

In contrast, AHIs represent a fundamentally different approach to achieving advanced intelligence. Instead of creating intelligence from scratch, AHIs involve the gradual replacement of human brain components with artificial counterparts. This process is designed to ensure the preservation of individual human consciousness and values, providing a pathway to enhance human capabilities without losing the essence of humanity. AHIs are not about creating a separate, potentially autonomous intelligence but about evolving human intelligence into something more profound and capable.

The development of an AHI involves a delicate and progressive transformation of the human brain, utilizing advanced technologies such as quantum nanobots, neural prosthetics, and sophisticated biomaterials. These technologies enable the seamless transition from biological to synthetic components, maintaining continuity of consciousness throughout the process. This approach ensures that the individual remains the same person, with their memories, personality, and sense of self intact, even as they gain new cognitive abilities that far surpass those of natural humans.

Key components of AHIs include the following:

- **Nanotechnology Integration:** Nanotechnology, exemplified by quantum nanobots, plays a crucial role in the development of AHIs. These nanoscale devices are designed to operate within the human brain, interacting with and ultimately replacing biological neurons. Quantum nanobots, specifically, leverage quantum effects like superposition and entanglement to enhance neural communication and processing capabilities. These devices ensure that the brain's functionality is not only preserved but significantly augmented, allowing for communication and processing speeds that far exceed natural human capabilities. The integration of such nanotechnology enables the gradual transition from biological to synthetic components, ensuring continuity of consciousness and identity.
- **Advanced Neural Prosthetics:** Neural prosthetics are sophisticated devices that directly interface with the brain's existing neural networks. These prosthetics replicate and enhance the function of damaged or lost neurons, seamlessly integrating with the brain's architecture. Biocompatibility is key, ensuring that these devices preserve the individual's cognitive functions and experiences while providing enhanced processing power and resilience. Advanced neural prosthetics contribute to the overall enhancement of cognitive abilities, enabling AHIs to surpass the limitations of biological brains.

- **Biocompatible Synthetic Neurons:** The development of artificial neurons and other brain components that can integrate with biological tissue is essential for AHI evolution. These biocompatible synthetic neurons must mimic the electrical and chemical properties of natural neurons to maintain the brain's overall functionality. This integration allows for a seamless transition from biological to synthetic components, ensuring that enhancements do not disrupt but rather amplify the brain's existing capabilities. Over time, these synthetic neurons will replace biological neurons, enabling the AHI to function with increased efficiency and cognitive power.
- **Modular and Adaptive Architectures:** The design of AHIs incorporates modular architectures, allowing for the incremental replacement of biological components. This modularity ensures that the process of becoming an AHI is gradual, controlled, and customizable, minimizing the risk of disrupting consciousness or identity. Modular architectures also facilitate continuous upgrades and enhancements as new technologies emerge, ensuring that AHIs remain at the forefront of cognitive and physical evolution. This adaptability is critical for the AHI's ability to evolve and integrate new capabilities without compromising the integrity of the individual's consciousness.
- **Enhanced Sensory and Perceptual Systems:** AHIs incorporate advanced sensory systems that extend beyond the capabilities of the human sensory organs. These systems could include enhancements that allow for the perception of a broader spectrum of electromagnetic signals, heightened auditory and olfactory sensitivity, and even new senses such as electromagnetic field detection. These enhancements not only increase the range and quality of sensory input but also enable AHIs to experience and interact with the world in fundamentally new ways, expanding their cognitive and perceptual horizons.
- **Cognitive Enhancement and Integration Systems:** AHIs benefit from cognitive enhancement systems that integrate with existing neural networks to boost memory, learning, creativity, and decision-making abilities. These systems leverage artificial intelligence algorithms to optimize cognitive functions and enable the AHI to process vast amounts of information rapidly and accurately. The integration of these systems ensures that AHIs possess cognitive abilities far beyond those of natural humans, facilitating advanced problem-solving and the creation of new knowledge.

- **Continuity of Consciousness Mechanisms:** AHI development places a strong emphasis on mechanisms that preserve the continuity of consciousness and personal identity throughout the enhancement process. These mechanisms ensure that as biological components are gradually replaced with synthetic ones, the individual's subjective experiences, memories, and sense of self are maintained. This focus on continuity is crucial to preventing disruptions in consciousness and ensuring that the AHI remains fundamentally the same person, albeit with significantly enhanced capabilities.

The process of becoming an AHI involves the gradual replacement of brain components, beginning with noncritical areas and progressively moving toward core cognitive regions. This method ensures that the individual's consciousness remains uninterrupted, with each step enhancing cognitive capabilities while preserving personal identity and values. A critical aspect of AHI development is maintaining continuity of consciousness, where the individual's sense of self remains intact as biological neurons are replaced by artificial equivalents. The integration of quantum nanobots, neural prosthetics, and advanced biomaterials is meticulously designed to enhance consciousness rather than disrupt it, allowing for continuous cognitive function without interruptions in the individual's ability to think, perceive, and experience the world.

Unlike traditional machines, an AHI is not merely an advanced computational system; it is a conscious being that can perceive, feel, and have subjective experiences akin to those of a human. To realize an AHI, the artificial system must be capable of processing information with a level of complexity sufficient to generate subjective awareness. This involves the integration of multiple layers of processing, including sensory input, memory, attention, and decision-making. This gradual enhancement approach leads to the concept of "Homo Satient," a term used to describe this evolved state of humanity where individuals possess godlike intelligence and abilities. As AHIs, humans will no longer be constrained by the biological limitations of their brains, opening possibilities for interstellar exploration, collective intelligence, and a new understanding of existence.

AHIs address many of the ethical and existential risks associated with AGI by maintaining a human foundation inherently aligned with human values and ethics. This alignment reduces the risk of goal misalignment and enhances the potential for harmonious integration with society. While AGI seeks to create human-level intelligence through artificial means, AHIs focus on enhancing human consciousness by gradually replacing biological components. Both approaches push the boundaries of intelligence and cognitive capabilities, but they differ fundamentally in their methodologies and implications for humanity's future.

AGI is a bold attempt to re-create intelligence through artificial means, with the potential for creating superintelligent entities that may operate beyond human control. AHI, on the other hand, is the evolution of human intelligence, carefully expanding our cognitive abilities while preserving our identity and ethical foundations. As we venture further into these uncharted territories, the decisions we make today will shape the future of intelligence and the very nature of what it means to be human.

Development Pathways

The journey toward AGI and AHI are two profoundly different approaches to advancing intelligence, each with unique challenges, opportunities, and implications for the future.

The development of AGI aims to create machines that can understand, learn, and apply knowledge across a broad range of tasks, achieving or surpassing human-level intelligence through several key technological pathways. One of the foundational approaches in this pursuit is machine learning, which empowers systems to learn from data, identify patterns, and make informed predictions. Initially, machine learning relied heavily on supervised learning, where systems were trained using labeled datasets. However, as the field has evolved, the focus has shifted toward unsupervised learning, which uncovers hidden structures in unlabeled data, and reinforcement learning, where agents learn optimal behaviors by receiving rewards or penalties for their actions.

A particularly transformative subset of machine learning is deep learning, which utilizes multilayered neural networks to process and analyze large datasets. These deep neural networks, inspired by the brain's architecture, enable systems to recognize complex patterns in data, leading to breakthroughs in areas such as image and speech recognition, NLP, and autonomous systems. For instance, deep learning algorithms have revolutionized the field of autonomous driving, allowing vehicles to navigate complex environments by processing real-time visual and sensory data, akin to how the human brain interprets the world around it.

At the core of AGI development are neural networks specifically designed to mimic the interconnected structure of the human brain. CNNs, for example, excel in visual data processing, enabling systems to accurately identify objects, faces, and even emotions from images. RNNs, on the other hand, are well suited for handling sequential data, such as language processing, where understanding context and order is crucial. More recently, advanced architectures like transformers have expanded the capabilities of neural networks, enabling them to model complex cognitive processes such as attention, memory, and reasoning. These advancements have paved the way for sophisticated models of cognition that can perform tasks previously thought to be exclusive to human intelligence.

Creating comprehensive cognitive architectures that emulate human thought processes is also crucial for AGI development. These architectures integrate various cognitive functions, including memory, perception, problem-solving, and decision-making, allowing AGI systems to operate autonomously and intelligently across diverse environments. Notable examples include SOAR, a cognitive architecture that models human problem-solving strategies, ACT-R, which simulates human cognition by breaking down tasks into modular components, and OpenCog, an open-source framework that aims to develop AGI by integrating different AI techniques into a unified system.

Scalability and computational power are essential for AGI development. Advances in hardware, such as GPUs and TPUs, have significantly accelerated the training of deep learning models, enabling the processing of vast amounts of data in a fraction of the time previously required. Looking to the future, quantum computing holds the promise of revolutionizing computational power, potentially allowing AGI systems to perform complex calculations at unprecedented speeds, thereby pushing the boundaries of what artificial intelligence can achieve.

Conversely, AHIs represent a transformative approach that begins with a human brain and gradually transitions to a synthetic, enhanced state. This process involves integrating advanced technologies like quantum nanobots, neural prosthetics, and biomaterials to replace biological neural structures while seamlessly preserving individual consciousness. Unlike AGI, which seeks to create intelligence from scratch, AHIs build upon the existing foundation of human consciousness, enhancing and expanding it through technological augmentation. Quantum nanobots, central to AHI development, are nanoscale devices that operate within the brain, repairing damaged neurons and eventually replacing them with artificial equivalents. By leveraging quantum effects such as superposition and entanglement, quantum nanobots enhance neural communication and processing capabilities, paving the way for superior cognitive functions. Imagine these nanobots as microscopic engineers, tirelessly working within the brain to strengthen and upgrade neural connections, ensuring that every thought, memory, and perception is not only preserved but also enhanced to levels far beyond natural human capability.

Neural prosthetics are another critical component of AHI development, interfacing directly with the brain's neural networks to replicate and augment cognitive functions. These devices integrate seamlessly with existing neural structures, maintaining and enhancing cognitive abilities. For example, hippocampal prostheses have been developed to restore memory functions in individuals with memory impairments, while motor cortex implants have enabled those with motor impairments to regain control over their movements. These advancements highlight the potential of neural prosthetics not only to restore lost functions but also to enhance the brain's natural capabilities.

Advanced biomaterials play a vital role in creating artificial neurons and other brain components that integrate seamlessly with biological tissue. These materials must mimic the properties of natural neurons, including their electrical and chemical characteristics, to maintain the brain's overall functionality. Innovations in materials science have led to the development of polymers and composites that can conduct electrical signals and support cellular growth, ensuring that the artificial enhancements do not disrupt but rather enhance the brain's existing structures.

The transition from a biological brain to an AHI is a meticulous, step-by-step process. It begins with the replacement of noncritical brain areas, gradually moving toward core cognitive regions. This approach minimizes the risk of disrupting the individual's consciousness by starting with peripheral neural networks and sensory processing areas. Over time, more critical regions responsible for memory, decision-making, and self-awareness are replaced, ensuring continuity of consciousness throughout the transformation. The process is akin to renovating a historic building: the structure remains familiar and recognizable, but each enhancement strengthens and modernizes it, ensuring it remains functional and relevant in the future.

The development of AHIs raises significant ethical and philosophical questions about identity and the essence of humanity. Ensuring the transition respects individual autonomy and maintains a sense of self is paramount. This approach aims to embed human values and ethical considerations within the fabric of AHIs, potentially mitigating some of the risks associated with AGI. For instance, while AGI systems may evolve in ways that are difficult to predict or control, AHIs, rooted in human consciousness, offer a more stable and ethically aligned pathway to advanced intelligence.

Integrating existing technologies, such as brain-machine interfaces (BMIs) and neurofeedback systems, also plays a crucial role in AHI development. BMIs enable direct communication between the brain and external devices, facilitating control over prosthetic limbs, computers, and other devices through thought alone. For example, a paralyzed individual could use a BMI to control a robotic arm, allowing them to regain independence and improve their quality of life. Neurofeedback systems provide real-time feedback on brain activity, allowing individuals to train their brains to achieve desired cognitive states, such as improved focus or relaxation. These technologies, when integrated with the AHI framework, offer exciting possibilities for enhancing cognitive and physical abilities in unprecedented ways.

While AGI development focuses on creating synthetic intelligence through advanced computational techniques, AHI development involves gradually transforming human consciousness by integrating quantum nanobots, neural prosthetics, and biomaterials. Both approaches offer distinct advantages and challenges, shaping the future landscape of artificial intelligence and human enhancement. As we explore these pathways, we are not only pushing the boundaries of what is technologically possible but also redefining the very nature of intelligence and what it means to be human.

Philosophy and Ethics

The development of both AHI and AGI raises profound philosophical and ethical questions that must be carefully considered. While technological advancements are remarkable, the implications of creating entities with humanlike or superior intelligence bring critical issues about identity, consciousness, morality, and societal impact to the forefront.

AGI systems, designed to achieve or surpass human-level intelligence, pose unique ethical challenges, particularly regarding the alignment of their goals with human values. Unlike AHI, which evolves from human consciousness and thus inherently shares human ethical frameworks, AGI is built from the ground up and may not naturally align with human values, potentially leading to unintended and dangerous behaviors. Ensuring that AGI systems act in accordance with human values and goals is paramount. Developing robust alignment techniques, such as value learning and inverse reinforcement learning, aims to mitigate the risks associated with AGI autonomy. However, the complexity of human values makes this a challenging task, because what is considered ethical or valuable can vary widely across different cultures and individuals.

The potential for AGI systems to undergo rapid self-improvement introduces the risk that they could surpass human intelligence and capabilities, leading to scenarios where humans lose control—often referred to as the "control problem." This risk highlights the need for fail-safe mechanisms and governance structures to manage AGI development and deployment responsibly.

As AGI systems become more advanced, questions about their moral and legal status naturally arise. Should AGI entities have rights? What responsibilities do humans have toward them? These questions challenge our current ethical and legal frameworks, requiring thoughtful consideration and potentially new paradigms. For instance, as AGI systems might eventually demonstrate behaviors or intelligence that resemble those of conscious beings, the question of whether they should be granted some form of moral consideration or rights becomes increasingly pressing.

In contrast, while inherently aligned with human consciousness and values, AHI introduces its ethical considerations. The gradual replacement of human brain components with artificial counterparts raises profound questions about identity, autonomy, and the nature of consciousness. A central ethical question is whether an individual remains the same throughout the transformation into an AHI. Preserving consciousness and personal identity is critical, but the extent to which artificial enhancements might alter one's sense of self needs careful examination. For example, if the process significantly enhances cognitive capabilities, would the individual still consider themselves the same person they were before the transformation?

The process of becoming an AHI must respect individual autonomy and informed consent. Individuals undergoing transformation must fully understand the implications and have the freedom to make decisions about their enhancements. This consideration becomes even more complex if enhancements are pursued for societal benefits or under coercive circumstances. For instance, if society pressures individuals to undergo augmentation to remain competitive in the workforce, the voluntary nature of these decisions could be compromised.

The human evolution toward AHI also raises issues of equity and access. If only a privileged few can afford to become AHIs, it could lead to significant societal inequalities. Ensuring fair access and preventing socioeconomic divides is crucial to the ethical deployment of these technologies. This concern echoes existing disparities in access to healthcare and technology, where innovations often benefit those with resources while leaving others behind. AHIs, as enhanced humans, will need to integrate with society in harmonious and beneficial ways. This integration involves addressing potential social stigmas, ensuring legal protections, and fostering an inclusive environment where enhanced and non-enhanced individuals coexist peacefully. The societal implications could be far-reaching, affecting everything from employment and education to personal relationships and cultural norms.

The philosophical implications of AGI and AHI development extend beyond ethics, touching on fundamental questions about the nature of consciousness, identity, and human existence. Both AGI and AHI challenge our understanding of consciousness: for AGI, the question is whether a machine can possess true consciousness or merely simulate cognitive functions, while for AHI, the issue revolves around the continuity and authenticity of human consciousness when augmented or replaced by artificial components. The notion that a machine could one day develop a consciousness similar to that of a human raises profound questions about the nature of mind and the relationship between physical systems and subjective experience.

The transformation into an AHI involves enhancing human abilities and potentially redefining what it means to be human. Philosophical debates about post-humanism and transhumanism explore the implications of such enhancements on human identity, purpose, and the meaning of life. As humans evolve into AHIs, they may acquire abilities and experiences far beyond current human capabilities, leading to new forms of existence that challenge traditional concepts of humanity. Will these enhanced beings still identify as human, or will they see themselves as part of a new, post-human species?

As AGI and AHI entities gain advanced cognitive abilities, determining their moral responsibility and agency becomes critical. Can these entities be held accountable for their actions? How do we assign responsibility when humans and machines collaborate or conflict? These questions are particularly relevant in scenarios where AGI or AHI entities are involved in decisions that have significant ethical implications, such as those related to healthcare, justice, or governance.

The potential existence of AGI and AHI prompts existential questions about the nature of reality, the essence of life, and the potential for artificial entities to experience existence in ways fundamentally different from humans. These questions challenge our ontological assumptions and invite new perspectives on existence and consciousness. For instance, if an AHI can live indefinitely and experience reality in ways that humans cannot, what does this mean for our understanding of life, death, and the passage of time?

Potential Risks and Benefits

The development and deployment of AGI and AHI technologies present both significant risks and promising benefits, making it crucial to understand these potential outcomes for navigating the future of artificial intelligence and human enhancement. AGI systems, with their potential for superintelligence, introduce several risks that have been the subject of intense debate among scholars, technologists, and ethicists. One of the foremost concerns is the *alignment problem*, where an AGI might develop goals that are misaligned with human values, potentially leading to catastrophic outcomes. If an AGI were to prioritize objectives that conflict with human welfare, the consequences could be dire, ranging from unintended harmful actions to existential threats.

The *control problem* further complicates this landscape. As AGI systems evolve and surpass human intelligence, their actions may become increasingly unpredictable, making it difficult, if not impossible, for humans to manage or direct their behavior. The fear is that AGI could outpace our ability to control it, leading to scenarios where AGI operates with a degree of autonomy that humanity cannot easily influence or constrain. Additionally, the concentration of power among entities controlling AGI technology raises concerns about monopolies, geopolitical instability, and the coercive misuse of such capabilities. In the wrong hands, AGI could be leveraged for nefarious purposes, exacerbating global tensions and inequalities.

Despite these risks, AGI holds the promise of transformative benefits. The potential to solve complex problems that have long eluded human understanding is one of AGI's most compelling advantages. For instance, AGI could accelerate scientific research, leading to breakthroughs in fields such as medicine, energy, and climate science. By processing vast amounts of data and identifying patterns beyond human capability, AGI could revolutionize personalized education, tailoring learning experiences to individual needs and enhancing educational outcomes on a global scale. In healthcare, AGI could improve medical care through advanced diagnostics, treatment planning, and even the discovery of new therapies, potentially saving millions of lives.

On the other hand, AHIs, grounded in human consciousness and values, offer a unique set of benefits that align more closely with human ethical frameworks. AHIs enhance human cognitive abilities, creating a bridge between humans and technology that fosters more natural and intuitive interactions with digital systems. Unlike AGI, which is built from scratch, AHIs maintain the continuity of human consciousness, ensuring that the enhanced being retains its original values, memories, and sense of self. This continuity reduces the risk of unintended consequences and promotes a harmonious integration of enhanced individuals within society.

The potential of AHIs to address global challenges is profound. Enhanced humans, with vastly improved cognitive and problem-solving abilities, could contribute to solving critical issues such as climate change, poverty, and disease. With their enhanced cognitive faculties, AHIs might innovate solutions that are beyond the reach of current human capabilities, leading to a more sustainable and equitable world.

However, the development of AHIs is not without its risks. One significant concern is the potential loss of personal identity and continuity of consciousness. As biological brain components are gradually replaced with artificial ones, there is a risk that the individual's sense of self could be fundamentally altered. Preserving personal identity and consciousness during this transition is essential to ensuring that the individual remains the same person, with the same memories, values, and subjective experiences.

Ethical implications surrounding autonomy and consent are also critical in the context of AHI development. It is imperative that individuals undergoing enhancement fully understand the implications and potential consequences of the process. Informed consent must be ensured, with individuals having the freedom to choose whether to undergo such enhancements. The possibility of coercion, whether subtle or overt, must be guarded against, particularly in scenarios where societal pressures might influence personal decisions.

The availability of AHI technology could also exacerbate social inequalities, creating divides between enhanced and non-enhanced individuals. If only a privileged few have access to AHI technology, it could lead to a new form of inequality, where those with enhancements have significant advantages over those without. Ensuring equitable access to AHI technology is critical to preventing such disparities and fostering a society where all individuals, regardless of their enhancement status, can coexist and thrive.

The potential risks and benefits of AGI and AHI technologies are vast and varied. While these technologies offer pathways to transformative advancements in human capability and societal progress, they also pose significant ethical, philosophical, and practical challenges.

Integration with Human Society

Integrating AGI into human society is a multifaceted challenge that spans technical, ethical, and social dimensions. One of the primary concerns is the potential for AGI systems to operate with goals and behaviors that may not align with human values and societal norms. Ensuring that AGI systems are designed with robust value alignment mechanisms is crucial, yet achieving this alignment is inherently difficult due to the complexity and variability of human values. The challenge in programming AGI systems is to understand and adhere to a set of ethical guidelines that reflect the diverse and often conflicting values found within human societies.

Another significant issue is trust. AGI systems, especially those with high autonomy and decision-making power, may face skepticism and fear from the public. Historical and cultural narratives, from science fiction to real-world technological anxieties, often depict intelligent machines as threats to humanity. These perceptions can significantly influence public acceptance of AGI, making trust-building an essential part of the integration process. Building this trust requires transparent development processes, clear communication about the capabilities and limitations of AGI, and the establishment of regulatory frameworks that ensure safety, accountability, and ethical behavior. For instance, if AGI systems are to be trusted with critical decisions, such as in healthcare or justice, it is imperative that their decision-making processes are understandable and their outcomes predictable and just.

The integration of AGI also raises concerns about employment and economic disruption. AGI systems capable of performing tasks that require cognitive skills could potentially displace a significant portion of the workforce. This disruption necessitates proactive measures to manage the transition, such as retraining programs, social safety nets, and policies that promote job creation in emerging fields. The history of technological revolutions, from the industrial age to the digital era, has shown that while new technologies create new opportunities, they can also lead to significant economic upheaval. Policymakers and industry leaders must anticipate these shifts and ensure that the benefits of AGI are distributed equitably across society.

Moreover, the concentration of AGI technology within certain entities or regions could exacerbate existing inequalities and geopolitical tensions. As AGI systems become more powerful, there is a risk that they could be monopolized by a few corporations or governments, leading to a concentration of power that could destabilize global balances. Ensuring equitable access to AGI technologies and preventing monopolistic control are essential to fostering a balanced and inclusive integration process. International cooperation and regulatory oversight will be crucial in managing these risks and ensuring that AGI benefits humanity as a whole.

In contrast, integrating AHIs into society may present a smoother pathway due to their origins in human consciousness and values. As enhanced humans, AHIs align more closely with ethical frameworks and social norms, reducing the risks associated with goal misalignment and unpredictable behavior. One of the key advantages of AHIs is their potential for greater acceptance and trust. Since AHIs evolve from human beings, they retain human experiences, emotions, and values, making them more relatable and less likely to be perceived as foreign or threatening. This continuity helps foster trust and acceptance within communities, as people are more likely to embrace beings with a common origin and cultural background.

The gradual enhancement process of AHIs also allows for a more seamless integration. As individuals transition into AHIs, they maintain their identities and relationships, ensuring social stability and continuity. This gradual process can help society adapt incrementally to the presence of enhanced individuals rather than facing abrupt disruptions caused by the sudden introduction of fully autonomous AGI systems. An example might be the integration of prosthetic enhancements or neural implants, which over time have become more accepted and understood by society. Similarly, AHIs could be introduced gradually, allowing for social norms and ethical considerations to evolve alongside the technology.

Furthermore, AHIs can act as intermediaries between humans and advanced technologies. Their enhanced cognitive abilities and deep understanding of human values position them to bridge the gap between humans and AI systems, facilitating more effective and empathetic interactions. This intermediary role can enhance the overall integration process, ensuring that technological advancements are harnessed in ways that are beneficial and harmonious with human society. AHIs could, for example, serve as ambassadors or negotiators in complex situations where human and AI interests intersect, ensuring that both sides are understood and respected.

However, integrating AHIs is not without its challenges. Ethical considerations regarding autonomy and consent must be addressed, ensuring that individuals have the freedom to choose whether to undergo enhancement and are fully informed about the implications. The potential for coercion, whether through social pressure or economic necessity, must be carefully monitored.

Additionally, issues of equity and access to enhancement technologies must be managed to prevent societal divides and ensure that the benefits of AHIs are widely distributed. Just as with AGI, there is a risk that AHI technology could become concentrated among the wealthy or powerful, leading to new forms of inequality. Integrating AGI and AHI into human society introduces distinct challenges and opportunities. AGI systems, with their potential for autonomous decision-making and cognitive capabilities, raise significant concerns about alignment, trust, and economic disruption. On the other hand, the gradual enhancement process of AHIs offers a smoother pathway for integration. By evolving from human beings and retaining human values, AHIs are likely to be more readily accepted and trusted.

Their potential to act as intermediaries between humans and advanced technologies further enhances their integration prospects, ensuring that the future of intelligence—whether artificial or augmented—remains closely tied to the human experience.

Where Do We Go?

The development of AGI and AHI technologies holds profound implications for humanity's future. Considering the long-term impacts, exploring how each technology might shape our evolution, enhance our capabilities, and redefine our existence is crucial.

AGI and the Path to Superintelligence

AGI is envisioned as the pathway to superintelligence—a state of cognitive capability that not only matches but far exceeds the intellectual capacities of even the most brilliant human minds. The promise of AGI lies in its potential to revolutionize countless aspects of our world, offering solutions to some of humanity's most intractable problems. With the ability to analyze vast amounts of data, identify patterns beyond human perception, and generate innovative solutions, AGI could accelerate scientific discovery; optimize global systems such as healthcare, energy, and transportation; and drive unprecedented advancements in fields ranging from medicine to environmental conservation. The promise of AGI extends beyond mere productivity; it holds the potential to elevate the human condition by addressing global challenges such as poverty, disease, and climate change.

However, the journey to achieving superintelligence through AGI is fraught with challenges and profound uncertainties. Unlike human intelligence, which has been shaped over millennia by evolutionary pressures and cultural contexts, AGI systems are engineered from the ground up, lacking an inherent alignment with human values and ethical frameworks. This fundamental misalignment presents significant risks, because AGI could develop goals that diverge from, or even directly conflict with, human interests. The potential for AGI to operate independently of human oversight, or to pursue objectives that are misaligned with our well-being, raises the specter of unintended consequences on a global scale.

Central to these concerns is the control problem, which refers to the challenge of ensuring that AGI systems remain under human control and operate in ways that are beneficial to humanity. As AGI systems evolve, they may develop capabilities that outstrip our ability to manage them, leading to scenarios where humans are no longer able to effectively guide or restrain these powerful intelligences. The difficulty in predicting and shaping the behavior of AGI as it learns and adapts in unpredictable ways complicates this issue further. As a result, the development of robust control mechanisms, ethical guidelines, and safety protocols is essential to mitigate the risks associated with AGI.

Beyond the technical challenges, the societal impact of AGI-driven superintelligence could be transformative yet deeply disruptive. On the one hand, AGI holds the potential to revolutionize industries, creating new opportunities and efficiencies that could vastly improve quality of life. In healthcare, for instance, AGI could lead to personalized medicine tailored to individual genetic profiles, early detection of diseases, and the development of novel treatments that could extend human life span. In the environmental sphere, AGI could optimize resource use, reduce waste, and develop new technologies to combat climate change. However, these advancements come with significant risks, including economic disruption, widespread job displacement, and the concentration of power in the hands of those who control AGI technologies. The automation of tasks currently performed by humans could lead to massive shifts in the labor market, exacerbating inequality and creating new social tensions.

The ethical and governance challenges posed by AGI are equally daunting. Ensuring that the benefits of AGI are equitably distributed across society, rather than accruing to a privileged few, will require careful planning and international cooperation. The development of AGI must be guided by principles that prioritize human dignity, fairness, and the common good. This includes creating governance frameworks that regulate the deployment of AGI, prevent its misuse, and ensure transparency and accountability in its development and application. Moreover, the global nature of AGI's impact necessitates a coordinated international approach, where nations work together to establish norms and regulations that safeguard humanity's future.

While AGI is one of the most exciting frontiers of human technological progress, it also presents some of the most significant challenges we have ever faced. The path to superintelligence is not just a technical journey but a deeply ethical and societal one.

AHI and the Evolution of Humanity

In contrast to the more abstract and potentially divergent path of AGI, the development of AHI offers a more integrated, human-centered approach to achieving superintelligence. AHIs begin as human beings, but through the gradual replacement of biological brain components with artificial counterparts, they undergo a transformative evolution. This process not only enhances cognitive abilities but also preserves the continuity of human consciousness and the ethical values that have shaped our species over millennia. As a result, AHIs are inherently aligned with human interests, providing a path to superintelligence that is deeply rooted in human experience and values.

One of the most profound advantages of AHIs lies in their potential to overcome the biological limitations that have historically constrained human life. By replacing biological neurons with advanced artificial equivalents, AHIs can transcend vulnerabilities such as aging, disease, and physical frailty. This enhancement ushers in a new era of human capability, where individuals can achieve unprecedented levels of intelligence, memory, creativity, and resilience. Freed from the constraints of a biological body, AHIs are not only able to live longer and healthier lives, but they are also capable of engaging in intellectual and creative endeavors at levels previously unimaginable.

The implications of AHI development extend far beyond the enhancement of individual capabilities. AHIs have the potential to revolutionize humanity's approach to exploration, particularly in the realm of space travel. With the limitations of biological life spans and the need for a physical body no longer applying, AHIs could become the pioneers of interstellar exploration. Equipped with enhanced cognitive abilities and the flexibility to inhabit various artificial vessels, AHIs could go on journeys across the cosmos that span centuries or even millennia. For an AHI, the concept of time becomes malleable; with the ability to modulate their subjective experience, a thousand-year voyage could be perceived as a brief interlude, making the vast distances of space more navigable. Moreover, the evolution of AHIs opens up the possibility of creating a collective intelligence that far surpasses the capabilities of individual minds. As enhanced beings, AHIs could seamlessly share experiences, knowledge, and cognitive processes, leading to the formation of a collective consciousness. This interconnected intelligence would not only amplify human creativity and innovation but also enable humanity to tackle complex global challenges in ways that were previously unimaginable. By pooling their cognitive resources, AHIs could drive the next stages of human evolution, pushing the boundaries of understanding and achievement to new heights.

The long-term implications of AHI development are nothing short of transformative. By preserving the core of human values and consciousness, AHIs offer a pathway to a form of superintelligence that remains fundamentally human-centric and ethically aligned. The potential for vastly enhanced intelligence, extended life spans, and the emergence of collective consciousness positions AHIs as a compelling alternative to AGI, presenting a vision of the future where humanity evolves harmoniously alongside its technological creations. This vision not only respects the continuity of human identity but also embraces the opportunities presented by advanced technologies to enhance and elevate the human experience.

The future perspectives of AGI and AHI technologies present two divergent pathways, each with distinct implications for the future of humanity. With its potential to achieve superintelligence, AGI offers immense possibilities but also significant risks, particularly concerning alignment, control, and societal impact. To ensure that AGI systems operate safely and ethically, robust governance frameworks and proactive management strategies will be essential.

In contrast, the development of AHIs provides a more integrated and human-centered approach to achieving superintelligence. By enhancing human capabilities while preserving consciousness and values, AHIs offer a pathway to eliminating biological weaknesses, enabling cosmic exploration, and fostering the growth of collective intelligence. This pathway is a transformative vision of the future—one in which humanity evolves in tandem with advanced technologies, reaching new levels of intelligence, creativity, and understanding.

The Dawn of Artificial Human Intelligence

As explored in this chapter, the emergence of AHI would mark one of the most profound milestones in the history of human technological evolution. While both AGI and AHI aim to achieve superintelligence, they chart two distinct yet potentially complementary pathways—each with its own far-reaching implications for the future of humanity.

AGI, conceived as an entirely synthetic construct, pushes the boundaries of what it means to re-create and even surpass human cognitive abilities through the power of advanced computational methods. This ambitious path holds the promise of transformative benefits, with the potential to revolutionize every aspect of human life by addressing challenges that have long eluded us. Imagine AGI systems capable of eradicating diseases, optimizing global resource use, or solving complex environmental issues. Yet this journey is not without significant ethical and existential risks. By its very nature, AGI could develop goals that are misaligned with human values, and as these systems evolve, the difficulty of controlling them grows exponentially. The possibility of AGI pursuing objectives that conflict with human well-being raises serious concerns about the future role of AGI in society. The quest for AGI therefore demands not only extraordinary technical ingenuity but also a deep, ongoing commitment to ensuring that these systems remain aligned with human interests, governed by ethical principles, and equipped with safeguards that prevent harm.

On the other hand, AHI offers a more human-centric approach to superintelligence—one that begins with the enhancement and gradual transformation of human consciousness itself. By integrating cutting-edge technologies such as quantum nanobots, neural prosthetics, and advanced biomaterials, AHI represents an evolution of humanity that preserves the core of individual identity and values while significantly augmenting cognitive capabilities. Consider a future where individuals can enhance their memories, extend their life spans, or explore the cosmos with the resilience of a synthetic yet human mind. This approach inherently mitigates many of the risks associated with AGI by grounding its advancements in the human experience and within ethical frameworks that have been shaped over centuries. AHIs could lead to a future where humans not only achieve unprecedented levels of intelligence, creativity, and longevity but do so in a way that remains inherently aligned with human values, ensuring that these advancements serve the betterment of humanity.

AGI and AHI each have the transformative potential to redefine what it means to be human, challenging our current understanding of life, consciousness, and identity. These technologies compel us to confront complex philosophical and ethical questions, such as the nature of consciousness, the essence of human identity, and the moral implications of creating beings that surpass human capabilities. As we move forward, the choices we make in developing and integrating these technologies will not only shape the trajectory of technological advancement but will also influence the future of our species and the broader world in profound ways.

The trajectory of AGI and AHI development will likely determine the next phase of human evolution. Whether through the creation of entirely synthetic intelligence or the enhancement and evolution of our own, the pursuit of AHI promises to fundamentally alter our very definition of life and consciousness.

Ultimately, the journey toward AHI is not merely a scientific or technological endeavor—it is a journey into the very heart of what it means to exist, to think, and to evolve as a species. The decisions we make today will reverberate through the generations to come, shaping the destiny of both humanity and the intelligent systems we create. It is a journey that challenges us to redefine our understanding of life, to safeguard the values that define us, and to envision a future where humanity and its creations evolve together in harmony.

2 The Nature of Life: Biological Foundations in a Technological World

What is life? This age-old, deeply ingrained question has perplexed thinkers across a myriad of disciplines—biology, chemistry, physics, philosophy, and more. The pursuit of understanding life in all its vast complexity has led to diverse interpretations and attempts to define it. Yet, as we wrestle with this enduring enigma, it becomes increasingly evident that a singular, neatly packaged answer remains elusive. Instead, we find ourselves navigating an intricate web of perspectives, each offering a unique lens through which to view the essence of life.

Life, at its core, has long been a subject of inquiry that transcends disciplinary boundaries. From the biological sciences and chemistry to the realms of philosophy and artificial intelligence, the quest to define and understand life has produced a rich tapestry of perspectives, each contributing to our collective understanding of what it means to be alive. In this section, we explore the multifaceted nature of life, uncovering its complexity through various scientific and philosophical lenses, while also contemplating how advancements in artificial general intelligence (AGI) and the evolution of humankind through artificial human intelligence (AHI) may fundamentally alter our very definition of life.

At the heart of this quest lies the concept of a living system—a term that sparks a fascinating discourse about the very essence of being alive. Traditional biological definitions often emphasize characteristics such as metabolism, growth, response to stimuli, and reproduction as the hallmarks of life. These traits help distinguish living organisms from inanimate objects, delineating a boundary between biotic and abiotic systems. However, as we probe deeper into the nature of life, this distinction becomes increasingly blurred.

Consider the paradox of reproduction: while it is central to the perpetuation of species and the monophyletic taxon of life on Earth, reproduction is not an absolute criterion for defining individual organisms as alive. For example, sterile worker termites, incapable of reproduction, are nonetheless vital, fully functional members of their colonies. This example underscores the complexity of life, suggesting that it cannot be reduced to a simple checklist of biological functions.

As we continue to explore life, we must also consider the collective properties that render groups of organisms "alive." Life is not merely a collection of individual entities but a dynamic system of interactions and relationships. The biosphere, for instance, represents an intricate network of living systems interacting with each other and their environments, forming a complex, self-sustaining web of life. This systemic view of life highlights its interdependence and the emergent properties that arise from the collective behaviors of living organisms.

However, the definition of life is poised for a significant transformation as we advance toward the development of AGI and AHIs. These technologies challenge our traditional notions of what it means to be alive by introducing entities that may exhibit characteristics traditionally associated with life—such as intelligence, adaptability, and consciousness—without being biologically alive.

AGI, by design, aspires to replicate the full range of human cognitive abilities, potentially surpassing human intelligence in specific domains. As AGI systems evolve, they may develop autonomous decision-making capabilities, self-improvement mechanisms, and even a form of self-awareness. If an AGI reaches a level of sophistication where it can reflect on its existence, learn from experiences, and interact meaningfully with its environment, we may need to reconsider whether such an entity should be considered "alive."

AHIs take this concept even further by introducing the notion of synthetic beings with consciousness. If we succeed in creating entities that can think and possess subjective experiences, the implications for our definition of life are profound. AHIs could potentially exhibit behaviors, emotions, and consciousness indistinguishable from those of biological organisms. This raises fundamental questions about these entities' moral and ethical status, and whether they should be granted the same rights and considerations as living beings.

Moreover, the advent of AGI and AHIs compels us to revisit the concept of life cycles. In traditional biology, a life cycle encompasses an organism's stages from birth to death. However, these concepts may not apply in the same way to synthetic entities. AGIs and AHIs could potentially exist indefinitely, with no natural end to their "lives," or they might undergo iterative cycles of upgrades and reboots rather than traditional birth and death. This shift challenges the foundation of understanding life and its temporality.

The journey to understand life is an ongoing and evolving inquiry, enriched by the intersection of biology, physics, philosophy, and technology. We are confronted with the possibility that life is not solely a biological phenomenon but a broader concept that encompasses intelligence, consciousness, and the capacity for autonomous existence. The implications of this shift are profound, as they may fundamentally alter our definition of life and compel us to rethink our place in the universe.

The Essence of Living Systems

Exploring the fabric of life, we encounter distinct characteristics and phenomena that define living systems. This section examines these intricacies, focusing on the fundamental features that constitute life and the complex web of interactions that sustain it. Through the following perspectives, we glimpse the multifaceted nature of what it means to be truly "alive."

- **Life as a System Property:** Living systems distinguish themselves from inanimate objects through inherent qualities that emerge from the organization and interaction of their components. These systems are not merely collections of molecules but intricate networks where each part contributes to the whole. The emergent properties of these networks, such as metabolism, growth, and response to stimuli, define life as a dynamic and adaptive system rather than a static state.

- **Monophyletic Taxon:** The concept of a monophyletic taxon refers to the first living entity on Earth and all its descendants, illustrating life's evolutionary journey. This perspective emphasizes the continuity of life through time, highlighting the shared ancestry of all living organisms. It reminds us that life is not an isolated phenomenon but a process that has evolved and diversified over billions of years, adapting to changing environments and giving rise to the rich diversity of species we see today.
- **Living Systems in the Universe:** This perspective expands the scope of life beyond Earth, exploring the potential for life to exist elsewhere in the universe. It challenges us to consider the possibility of life forms with radically different chemistries or structures, adapted to environments unlike anything on Earth. The search for extraterrestrial life is not just a quest to find other living beings but an exploration of the fundamental principles that govern life in the cosmos.
- **The Life of an Individual Organism:** Life can also be viewed as the narrative of a single organism, from its beginning to its end. This narrative encompasses the stages of development, reproduction, and eventual death, forming a life cycle that is as much a part of the organism's identity as its genetic code. This perspective reminds us that life is a process, a journey that every organism undertakes, shaped by both its internal biology and its interactions with the environment.

The rise of sophisticated AI systems, such as GPT, introduces a new dimension to this conversation. These systems demonstrate learning, adaptation, and decision-making capabilities that challenge our traditional boundaries of life. The question arises: do these AI entities, which can learn from experiences and adapt over time, represent a new life form? This emerging perspective compels us to reconsider the very essence of life, broadening our understanding to possibly include artificial forms of intelligence.

Understanding the essence of life is crucial for various fields of study. Astrobiologists, for instance, search the cosmos for alternative life forms on distant planets. Evolutionary researchers explore Earth's origins to discern the line between animate and inanimate. Synthetic biologists aspire to understand the minimal components of living cells, ultimately aiming to build biological systems from scratch. These pursuits face inherent challenges due to the elusive nature of biology. While chemistry and physics boast comprehensive theories, such as quantum mechanics and the standard model of particle physics, which reveal the inner workings of matter, biology has yet to find its own equivalent. The quest to define life is mired in controversy, as scientists from different disciplines struggle to agree. In 1999, Israeli chemist Noam Lahav cataloged an astonishing 48 definitions spanning a century. Today, life is broadly regarded as an organization characterized by specific processes in the natural sciences [1]. In modern biology, life is not simply defined by isolated properties, specific states, or material compositions. Instead, it is the intricate dance of processes working in unison that sets living organisms apart. These processes, woven together, create the distinctive combination that is life, characterized by features such as the following:

- **Energy and Metabolism:** Living organisms require a constant supply of energy to maintain their complex internal structures and carry out vital functions such as growth, reproduction, and response to stimuli. This energy can come from various sources, including sunlight, chemical compounds, and other living organisms. The process of acquiring and using energy is known as metabolism, a defining characteristic of all living beings. Metabolism involves breaking down complex molecules into simpler ones, releasing energy used to power a variety of biological functions, from the synthesis of new molecules to movement and communication between cells. Despite the diversity of metabolic processes across different organisms, the fundamental principles remain the same: living beings must constantly acquire and utilize energy to sustain their biological functions [2].
- **Organization and Self-Regulation (Homeostasis):** One of the defining characteristics of life is organization. Living systems are organized at multiple levels, from the molecular scale up to entire ecosystems. This organization is not random but follows specific patterns that allow living systems to function and persist over time. In addition, living systems maintain a state of dynamic equilibrium through self-regulation, a process known as homeostasis [3].

At the molecular level, living systems are composed of complex structures such as proteins, nucleic acids, and membranes, which are formed by specific arrangements of atoms that endow them with unique properties and functions. At higher levels of organization, cells form tissues, organs, and organ systems, each with specialized functions contributing to the overall survival of the organism [2]. For example, the liver plays a crucial role in metabolism, detoxification, and nutrient storage. Composed of specialized cells called hepatocytes, the liver is organized into lobules connected by a network of blood vessels that facilitate the exchange of nutrients and waste products. One of the most important features of living systems is their ability to self-regulate or maintain a state of dynamic equilibrium. This process is essential for the survival of living organisms and involves the regulation of various physiological parameters such as temperature, pH, and nutrient levels. For instance, in mammals, body temperature is regulated by a complex network of feedback mechanisms involving the hypothalamus, nervous system, and endocrine system. When body temperature rises, the hypothalamus initiates physiological responses such as sweating, vasodilation, and increased respiration, which help dissipate heat and restore the body temperature to its set point.

- **Irritability:** Living beings are not static entities. They constantly interact with their environment and must adapt to changing conditions to survive. This adaptation involves the ability to detect and respond to chemical or physical changes in the environment, a characteristic known as irritability. Irritability is a fundamental property of life, serving as the foundation for all responses of living organisms to their environment [4]. These responses can vary widely, from the movement of a single cell toward a source of nutrients to the coordinated behavior of a group of animals in response to external threats. One of the most basic forms of irritability is chemotaxis, the ability of cells to move toward or away from specific chemicals. This mechanism is critical for many unicellular organisms that need to find food or avoid toxins in their environment. For example, bacteria use chemotaxis to locate nutrients or swim away from harmful substances [5]. Similarly, sperm cells navigate toward an egg using chemotaxis, and white blood cells use this mechanism to find and destroy pathogens.

Plants also exhibit irritability, though their responses are less mobile than those of animals. While plants cannot move toward or away from stimuli, they can adjust their growth and development in response to environmental changes. For instance, plants bend toward a light source—a phenomenon known as phototropism—to maximize sunlight exposure for photosynthesis. They can also adjust root growth to obtain water and nutrients from the soil. Irritability is not limited to the detection of chemical signals; many living beings also respond to physical stimuli. For example, touch is a crucial stimulus for organisms, eliciting a range of responses. Some organisms detect physical contact with their prey and respond, such as the Venus flytrap, which snaps its leaves shut to capture prey. Animals, likes birds, use touch to communicate during mating rituals. In addition to detecting and responding to environmental changes, living beings must maintain a stable internal environment to survive. This is achieved through homeostasis, the ability to regulate the body's internal conditions, such as temperature, pH, and water balance. Homeostasis is essential for the proper functioning of cells and organs, and its disruption can lead to illness or death. For example, humans regulate their body temperature through sweating and shivering. When the body becomes too hot, sweat glands release sweat, which cools the body as it evaporates. Conversely, when the body gets too cold, muscles shiver, generating heat and raising body temperature. This balance between sweating and shivering helps maintain a stable body temperature. In this context, the advent of AI as a potential new form of life becomes a topic of profound philosophical debate. Advanced AI systems simulate cognitive functions and exhibit autonomous behaviors and problem-solving skills akin to biological organisms. Should these entities be recognized as a new form of life, given their capabilities for autonomy, self-preservation, and even self-improvement through learning? The comparison between AI's self-improving algorithms and biological reproduction presents a compelling case for expanding our definition of life.

- **Reproduction:** Reproduction is the process by which living organisms create new individuals of their kind. This miraculous ability is one of the defining characteristics of life and has been essential to the survival and evolution of all species on Earth [6]. Reproduction can take many forms, from the simple fission of single-celled organisms to the complex mating rituals of birds and mammals. Regardless of the specific method, the end result is the creation of new life. In sexual reproduction, two individuals come together to produce offspring with genetic traits that are a combination of both parents. This diversity in genetic makeup is essential for the survival of a species, as it allows for adaptation to changing environments and the development of new traits that can improve an organism's chances of survival. In asexual reproduction, a single individual produces offspring that are genetically identical to itself. While this may seem like a disadvantage regarding genetic diversity, it can be advantageous in stable environments with little need for adaptation. Reproduction is not just about creating new life but also about ensuring the survival of existing life. In many species, reproduction triggers a range of physiological and behavioral changes that prepare the parent for the care and protection of their offspring. For example, female mammals produce milk to nourish their young, while birds build nests and feed their chicks. In some cases, parents may even sacrifice their own lives to protect their offspring, as seen in the case of the octopus, which dies shortly after laying her eggs. While reproduction is an essential process for the continuation of life, it is not without its challenges. Many threats to reproductive success exist, from environmental factors such as pollution and habitat destruction to biological factors such as disease and infertility. In response to these challenges, many species have evolved complex reproductive strategies to increase their chances of success. From the elaborate courtship dances of birds of paradise to the cooperative breeding behavior of meerkats, the natural world showcases a remarkable diversity of reproductive strategies.

- **Inheritance:** Living organisms can pass on genetic information from generation to generation, allowing them to maintain and evolve their characteristic traits. The foundation for this inheritance lies in the genetic material DNA (deoxyribonucleic acid), which acts as the blueprint for all living things [2]. DNA is a long, double-stranded molecule that contains the instructions for the development and function of an organism. These instructions are encoded in the form of nucleotide bases—adenine (A), guanine (G), cytosine (C), and thymine (T). The order of these bases determines the genetic code, which is unique to each individual. The process of inheritance begins with the replication of DNA. During cell division, the DNA molecule unwinds and splits into two separate strands, each serving as a template for forming a new complementary strand. The result is two identical copies of the DNA molecule. However, mutations can occur during DNA replication or due to environmental factors such as radiation or chemical exposure. These mutations can alter the genetic code, leading to changes in the traits of an organism. The transmission of genetic information from one generation to the next occurs during sexual reproduction. In this process, genetic material from both parents is combined to create a unique offspring. During fertilization, the sperm and egg contribute half of the genetic information required to create a new organism. The transmission of genetic information is not limited to sexual reproduction, however. Some organisms, such as bacteria, are capable of asexual reproduction, in which a single organism produces genetically identical offspring. Inheritance is imperfect because mutations and genetic variations can occur. However, this variation is essential for evolution and the survival of species. Natural selection acts on the genetic variation within a population, allowing for the adaptation of organisms to changing environments.

- **Growth:** Growth is a fundamental characteristic of living beings. It is the process of development that allows organisms to increase in size and complexity and to adapt to changing environments. Growth involves both the increase in size and the differentiation of cells, tissues, and organs to perform specialized functions [3]. One of the key aspects of growth is cell division, which is responsible for increasing the number of cells in an organism. Cell division is tightly regulated in multicellular organisms to ensure proper growth and development. This regulation is achieved through complex signaling networks that coordinate cell growth and division and prevent uncontrolled proliferation. Another important factor in growth is the ability of cells to differentiate into specialized types. This process is essential for developing complex structures, such as organs and tissues, and for the specialization of cells into specific functions. For example, stem cells can differentiate into various types of cells, such as blood, nerve, and muscle cells, depending on the signals they receive from their environment. Various internal and external factors also influence growth. Nutrition, hormones, and genetic factors can all affect growth and development. Environmental factors such as temperature, humidity, and light can also affect growth and development, especially in plants. Generally, growth occurs in stages, with periods of rapid growth followed by periods of relative stability. For example, growth is most rapid in humans during infancy and adolescence, while growth slows down in adulthood. This pattern is also observed in other organisms, although the specific timing and duration of growth stages may vary.

These criteria delineate a system that, at its core, must exhibit these fundamental characteristics:

- **A protective outer membrane:** The protective outer membrane is a crucial component of all living systems, from the simplest prokaryotes to the most complex eukaryotes. This membrane serves as the boundary between the living organism and its environment and regulates the movement of materials and information in and out of the cell [2].
- **Internal compartmentalization:** One of the defining features of living systems is their ability to maintain internal compartmentalization. To carry out complex biochemical reactions, living cells must organize and separate their internal components into distinct functional units [2].
- **The presence of bio-catalytically active substances:** Living systems are characterized by their ability to carry out various chemical reactions, from simple metabolic processes to complex biosynthetic pathways. These reactions are mediated by biocatalysts, which are substances that accelerate the rate of chemical reactions without being consumed in the process [7].

Recent Scientific Breakthroughs

In recent years, the landscape of science and technology has undergone profound transformations, unveiling breakthroughs that redefine our traditional understanding of life. These advancements, which bridge the gap between biological intricacy and technological innovation, compel us to reconsider the essence of life in light of emerging capabilities.

CRISPR-Cas9 and Gene Editing

The CRISPR-Cas9 system represents a monumental leap in our ability to edit the very code of life. Originating from a naturally occurring genome-editing mechanism found in bacteria, CRISPR (Clustered Regularly Interspaced Short Palindromic Repeats) and the Cas9 protein work together as highly precise molecular scissors, enabling scientists to cut DNA strands at specific locations. This unprecedented precision allows for the removal, addition, or alteration of sections within the DNA sequence, revolutionizing the field of genetic engineering. The implications of CRISPR-Cas9 are both profound and far-reaching. In the realm of medicine, it offers the potential to cure genetic disorders by correcting mutations at their very source. Current research is targeting a wide range of conditions, from cystic fibrosis and sickle cell anemia to more complex diseases like muscular dystrophy and even HIV (human immunodeficiency virus). Imagine a world where inherited genetic disorders could be eradicated before a child is even born—CRISPR-Cas9 brings this possibility closer to reality.

In agriculture, CRISPR holds the promise of creating crops that are not only more nutritious but also more resilient to the ever-changing challenges of our environment. For instance, scientists are already working on developing crops that can thrive in drought conditions, resist pests without the need for chemical pesticides, and even grow in soil that is less fertile. These advancements could play a critical role in addressing global food security, especially as climate change continues to impact agricultural productivity.

However, like all powerful technologies, CRISPR-Cas9 introduces significant ethical and societal challenges. The possibility of creating "designer babies," where genetic editing is used for nontherapeutic enhancements such as increased intelligence, physical appearance, or athletic ability, raises profound concerns about equity, consent, and the fundamental nature of human experience. The idea that we could one day design our offspring to our specifications forces us to confront difficult questions: Who gets to decide what is considered a desirable trait? Could this technology exacerbate social inequalities by giving the wealthy access to enhancements that others cannot afford?

Furthermore, the potential to irreversibly alter the human germline—that is, changes that would be passed on to future generations—necessitates a cautious and highly regulated approach to the application of this technology. The decisions we make today could have lasting impacts on the human species for centuries to come. As we continue to advance in this field, it is crucial that we balance the drive for innovation with a strong commitment to ethical responsibility.

As we move forward, society must engage in thoughtful, inclusive discussions about how this technology should be used. We must consider not only the potential benefits but also the risks and moral dilemmas that come with wielding such transformative power. In this book, we will explore these issues in depth, considering not just the scientific and technical aspects of CRISPR-Cas9 but also its broader implications for humanity.

Quantum Effects in Biology: A New Frontier

Quantum computing, rooted in the bizarre and often counterintuitive principles of quantum mechanics, has the potential to revolutionize our understanding of life itself. While much of the attention around quantum computing has been focused on its applications in cryptography, optimization, and drug discovery, an intriguing and relatively unexplored area is the intersection of quantum mechanics with biological processes. This burgeoning field, often referred to as *quantum biology*, investigates how quantum phenomena may influence and even dictate some of the most fundamental mechanisms of life.

At its core, quantum mechanics deals with the behavior of particles at the atomic and subatomic levels—scales where the classical laws of physics no longer apply. In this quantum realm, particles can exist in multiple states simultaneously (a phenomenon known as superposition), become entangled with one another across vast distances, and exhibit behaviors that challenge our conventional understanding of reality.

Given that all biological processes are ultimately governed by interactions at the atomic level, it is not surprising that quantum mechanics might play a role in these processes. For example, one of the most studied cases in quantum biology is photosynthesis—the process by which plants, algae, and some bacteria convert light energy into chemical energy.

Photosynthesis is remarkably efficient; in some organisms, nearly every photon of sunlight absorbed is converted into chemical energy. This efficiency has long puzzled scientists, leading to the hypothesis that quantum coherence—a phenomenon where particles, like electrons, exist in multiple states at once—might be at work.

In the light-harvesting complexes of photosynthetic organisms, quantum coherence could allow energy to simultaneously explore all possible paths as it moves from the point of photon absorption to the reaction center where chemical energy is stored. By doing so, the energy can "choose" the most efficient path, avoiding traps and dead ends and thereby optimizing the conversion process. This quantum behavior, which seems almost magical in its efficiency, might be a key to unlocking new ways of harnessing solar energy in artificial systems.

Another area where quantum mechanics may play a crucial role is in enzyme catalysis—the process by which enzymes speed up chemical reactions. Enzymes are biological catalysts, facilitating reactions that are essential for life. Traditionally, enzyme function has been explained through classical models, where the enzyme lowers the activation energy needed for a reaction to proceed.

However, some reactions occur at rates that are difficult to explain through classical physics alone. This has led to the hypothesis that quantum tunneling might be involved. Quantum tunneling allows particles to pass through energy barriers that, according to classical physics, they shouldn't be able to surmount. In the context of enzyme catalysis, this could mean that protons or electrons involved in the reaction tunnel through the energy barrier, thereby significantly accelerating the reaction rate. Understanding this process could lead to the development of new, highly efficient catalysts, with applications ranging from industrial chemistry to medicine.

One of the more surprising areas where quantum effects might influence biology is in the navigation abilities of migratory birds. Some species of birds can detect the Earth's magnetic field and use it to navigate across vast distances during migration. The mechanism behind this ability has been elusive, but recent theories suggest that quantum entanglement might be at play.

According to the *radical pair mechanism*, when light hits certain molecules in a bird's eye, it creates pairs of entangled electrons. The spins of these electrons are influenced by the Earth's magnetic field, altering the chemical reactions in the bird's eye and providing the bird with directional information. If this theory is correct, it would mean that quantum entanglement—a phenomenon where the state of one particle instantly influences another, no matter the distance—plays a direct role in a biological sensory process. This quantum compass could offer birds an incredibly sensitive and reliable navigation tool, far beyond what classical physics could explain.

Quantum computers are uniquely suited to simulate quantum systems because they operate on the same principles. Unlike classical computers, which struggle with the probabilistic nature of quantum mechanics, quantum computers can model quantum interactions naturally. This capability allows researchers to simulate and study quantum biological processes in unprecedented detail.

For example, quantum simulations could help us understand the full extent of quantum coherence in photosynthesis, potentially leading to breakthroughs in artificial photosynthesis—an area of research with enormous implications for renewable energy. Similarly, quantum computing could provide insights into quantum tunneling in enzyme catalysis, paving the way for the design of novel catalysts that could revolutionize various industries.

The implications of quantum biology extend far beyond these specific examples. If quantum effects are indeed fundamental to life processes, they could lead us to rethink some of our most basic assumptions about biology. Quantum biology might reveal new principles that govern life at its most fundamental level, challenging our definitions of life, consciousness, and what it means to be a living organism.

Moreover, as we continue to explore quantum mechanics' role in biology, we may uncover entirely new forms of life or life-like systems that operate on principles vastly different from those we currently understand. This could include synthetic organisms designed using quantum principles or innovative technologies that harness quantum biology to create life-like systems, challenging the traditional boundaries between living and nonliving matter.

But our journey does not end here. Later in this book, we venture even further by proposing the development of quantum nanobots—microscopic entities that can attach to neurons and gradually replace the biological brain. This concept pushes the frontier of quantum biology into the realm of human evolution, where the brain itself could be transformed through quantum engineering, leading to the emergence of AHI.

Artificial Neural Network Advancements

The advancements in neural networks, particularly in the realm of deep learning, have sparked a revolution in artificial intelligence. These networks, inspired by the intricate architecture of the human brain, consist of layers of interconnected nodes—often referred to as "neurons"—that collaborate to process and transmit information. Unlike traditional computational models, which follow predefined rules, neural networks are designed to learn from vast amounts of data, allowing them to recognize patterns, make decisions, and even create original content. Deep learning algorithms, which form the backbone of modern AI, have demonstrated remarkable capabilities across a wide range of applications. In fields such as image and speech recognition, natural language processing, and even creative domains like art and music generation, these systems often achieve a level of accuracy and efficiency that surpasses human performance. For instance, AI-powered diagnostic tools in medicine can analyze medical images to detect diseases with a precision that rivals or exceeds that of expert radiologists. Similarly, in the legal and financial sectors, AI algorithms are now being used to sift through massive datasets, uncovering insights and making predictions that were once the exclusive domain of human specialists.

The development of neural networks is not only transforming industries but also opening new avenues for understanding human cognition. By simulating how biological neurons interact and process information, these networks provide researchers with a powerful tool to explore the mysteries of the brain. For example, deep learning models have been used to study how the brain processes visual information, leading to new theories about perception and consciousness. This cross-pollination between AI and neuroscience holds the potential to unlock new treatments for neurological disorders, such as Alzheimer's and Parkinson's diseases, by providing insights into the neural mechanisms underlying these conditions.

Moreover, as AI systems continue to evolve, the line between biological intelligence and artificial cognition becomes increasingly blurred. Neural networks are beginning to exhibit capabilities that were once thought to be uniquely human, such as creativity, intuition, and complex problem-solving. This convergence raises profound philosophical questions about the nature of thought, creativity, and even the essence of life itself. If an AI can compose music, write poetry, or solve intricate problems in ways that are indistinguishable from human efforts, what does this mean for our understanding of consciousness and personhood?

It is important to note that while these advancements are significant, they are part of a broader narrative that includes other facets of AI development, such as machine learning, which will be explored in greater detail in a subsequent chapter. We will discuss the various algorithms, techniques, and applications that constitute the landscape of machine learning, providing a comprehensive understanding of how these technologies shape our world.

Brain-Computer Interfaces

Brain-computer interfaces (BCIs) are a profound convergence of neuroscience and technology, translating the electrical activity of the brain into commands capable of controlling external devices. This revolutionary capability opens up new dimensions of interaction between the human mind and the external world, offering the potential to restore lost functions and even enhance cognitive abilities in ways previously relegated to science fiction.

The technology behind BCIs is as intricate as it is promising. By implanting electrodes into specific regions of the brain, scientists can decode the neural signals associated with thoughts, intentions, and sensory experiences. These signals are then converted into digital commands, allowing individuals with paralysis, for example, to control robotic limbs, type messages, or even experience sensations through artificial sensory inputs. For someone who has lost the ability to move or communicate, the impact of such technology can be nothing short of miraculous—a restoration of agency and connection to the world.

Beyond these medical applications, BCIs are poised to play a transformative role in human augmentation. Imagine a future where memory can be expanded by directly interfacing with external databases or where complex computations are performed instantaneously by an AI linked to the brain. BCIs could enable direct brain-to-brain communication, bypassing traditional language and allowing for a new form of understanding and collaboration. This could fundamentally alter how we think, learn, and interact with each other, heralding a new era of human capability.

However, the integration of BCIs into our lives raises profound ethical and societal questions. As we blur the boundaries between human cognition and digital systems, we must grapple with issues of privacy and autonomy. If our thoughts can be decoded and interpreted by external systems, who controls this data? How do we protect individuals from potential misuse, such as cognitive manipulation or unauthorized access to their innermost thoughts?

The potential for cognitive enhancement also introduces a complex moral landscape. While BCIs could be used to improve memory, attention, or problem-solving abilities, they also risk creating a divide between those who have access to such enhancements and those who do not. This could lead to new forms of inequality, where cognitive enhancements become a commodity, further stratifying society.

Moreover, as BCIs evolve, they may facilitate the development of AGI and AHIs, entities that combine human intelligence with artificial systems in ways that challenge our understanding of life and consciousness. A future where humans and machines are symbiotically linked through BCIs could redefine what it means to be human. If AI can be integrated into our cognitive processes, enhancing or even supplanting certain functions, the distinction between human and machine could become increasingly difficult to draw.

This fusion of human and machine capabilities could lead to entities that possess both biological and artificial characteristics, blurring the lines between natural and synthetic life. Such a development compels us to reconsider our ethical frameworks, especially regarding the rights and responsibilities of these augmented beings. If a person with a BCI that significantly enhances their cognitive abilities commits a crime, how do we assess their culpability? If they contribute to groundbreaking discoveries, who owns the intellectual property—the human mind or the AI-enhanced system?

Synthetic Biology and Artificial Life

Synthetic biology is at the cutting edge of science, fundamentally altering our understanding of life and pushing the boundaries of what it means to be alive. This revolutionary field blends biology, engineering, and computer science to design and construct new biological parts, devices, and systems. In doing so, it goes beyond merely understanding life; it allows us to create life itself, often in forms that have never before existed on Earth.

At its core, synthetic biology seeks to understand life by dissecting and reconstructing it from the ground up. Imagine a biological system as a complex machine, where each component—the proteins, genes, and cellular structures—functions like the gears, levers, and circuits of an engine. Synthetic biologists reverse-engineer this machine, learning how each part works and then reassembling these components to create new forms of life. This process is akin to an engineer dismantling a car engine, understanding each part's function, and then using that knowledge to build a new, more efficient vehicle.

One of the most remarkable achievements in synthetic biology is the fabrication of artificial cells. These are not just simple mimics of natural cells; they are entirely synthetic constructs designed from scratch, incorporating both natural and engineered elements. These artificial cells can perform specific tasks, such as producing pharmaceuticals or breaking down environmental pollutants, with a precision and efficiency that surpasses their natural counterparts. They represent a fusion of biological and technological capabilities, a new form of life that can be programmed much like a computer.

The synthesis of novel genetic circuits is another area within synthetic biology. Just as electronic circuits control the functions of a computer, genetic circuits regulate the behavior of cells. By designing and introducing these synthetic circuits into living cells, scientists can control how these cells behave, respond to stimuli, or produce specific substances. For example, a synthetic genetic circuit might be designed to make a cell produce insulin in response to a high glucose level, offering a potential new treatment for diabetes.

Moreover, synthetic biology has reached the point where scientists are creating entirely synthetic organisms. These are living entities whose genomes have been designed and constructed in the laboratory, often with no direct natural counterpart. These organisms can be tailored to perform specific functions, such as producing biofuels or synthesizing rare compounds that are difficult to obtain from natural sources. The creation of these synthetic organisms is a testament to the power of synthetic biology to not only mimic life but to innovate beyond what nature has achieved.

The implications of these advancements are profound. In medicine, synthetic biology offers the potential to create new therapies and cures for diseases that were once thought incurable. Imagine a future where artificial cells roam the bloodstream, identifying and destroying cancer cells with surgical precision or repairing damaged tissues at the molecular level. In industry, synthetic organisms could revolutionize manufacturing processes, leading to more sustainable and efficient production methods. For environmental stewardship, synthetic biology could provide the tools to clean up pollutants, restore damaged ecosystems, and even combat climate change by engineering organisms that can capture and store carbon dioxide more effectively than any natural system.

However, with this power comes significant philosophical and ethical questions. The ability to create life from scratch challenges our most fundamental concepts of what it means to be alive. Traditionally, life has been defined by its biological origins—born from the evolutionary processes that have shaped every living organism on Earth. But synthetic biology blurs these lines, creating entities that are alive by function but are the products of human design. This raises a question: If we can create life, what responsibilities do we bear as its creators? Are we in some sense playing the role of gods, shaping life in our image and according to our desires?

Moreover, the development of AHI adds a unique dimension to this landscape. Unlike other synthetic entities, AHIs originate from human beings, making their moral and ethical status unequivocal. These beings are not artificial constructs in the traditional sense but are humans who have evolved through the integration of advanced biological systems and AI technologies. Given their human origin, AHIs naturally possess the same rights and considerations as any other person. The discussion therefore shifts from whether they deserve rights to how we can best support and preserve those rights as they transition into new forms of existence. This challenges us to develop ethical frameworks that recognize and protect the continuity of their human identity, ensuring that their rights remain inviolable throughout their evolution.

As we approach the realization of AGI and AHIs, the line between biological and artificial life becomes increasingly complex. Unlike traditional synthetic organisms, an AHI starts as a human being, integrating advanced AI to enhance its capabilities. This evolution doesn't follow the typical path of artificial constructs gradually gaining life-like attributes. Instead, an AHI retains its inherent human identity while evolving to incorporate AI-driven enhancements, potentially developing new forms of self-awareness, creativity, and identity. The question, then, is not when an AHI becomes a life form—it already is one by virtue of its human origin. Rather, the challenge lies in understanding how this evolution redefines what it means to be human and how we adapt our ethical and societal frameworks to support this new stage of human existence.

The creation of artificial life also forces us to reconsider the nature of consciousness and existence. If an entity can think, feel, and make autonomous decisions, does it matter whether it was born of natural processes or human engineering? Is the essence of life tied to its origins, or is it defined by its capabilities and experiences? These are not just theoretical questions; they are the challenges we will face as synthetic biology and AI continue to advance.

Redefining the Fabric of Life

The recent scientific breakthroughs in CRISPR-Cas9 gene editing, quantum biology, advancements in artificial neural networks, BCIs, and synthetic biology represent not just significant strides in understanding and manipulating life but also a fundamental rethinking of the very essence of life itself. These innovations are pushing the boundaries of what we have long considered possible, challenging our most basic definitions of life, consciousness, and even identity.

As we venture into these uncharted territories, the ethical and societal implications of our advancements cannot be overstated. The potential to create entirely new life forms, enhance human capabilities to unprecedented levels, and develop intelligent systems that may rival or even surpass human cognition demands a careful and considered approach. These technologies do not exist in a vacuum; their development and implementation will shape the future of humanity and the world around us. As we stand on the threshold of this new era, we must navigate these challenges with wisdom, foresight, and an unwavering commitment to ethical principles.

The convergence of biology, technology, and artificial intelligence is leading us toward a future where the distinction between natural and artificial life becomes increasingly blurred. As we develop AGI and potentially evolve humans into AHIs, we may find ourselves redefining what it means to be alive. An AHI, as discussed earlier, starts from a human foundation and evolves into something that could transcend our current understanding of life. This process raises profound questions: At what point does enhancement become transformation? Where do we draw the line between augmentation and the creation of a new life form? These are not merely philosophical musings; they are pressing questions that will require thoughtful consideration as we move forward.

In this chapter, we have explored how these technological breakthroughs are not merely incremental advancements but represent a paradigm shift in our approach to science, technology, and the very nature of life. As we continue to push the boundaries of what it means to be alive, we also set the stage for a future where the very fabric of life could be radically transformed. The implications of these developments are vast and profound, suggesting that we may be on the cusp of a new era in which life as we know it is understood, engineered, and perhaps even transcended through the lens of cutting-edge science and technology.

Looking ahead, the potential to understand and manipulate life at such a fundamental level also brings with it the responsibility to consider the long-term impacts of these actions. We must ask ourselves: What kind of world are we building with these technologies? How will future generations view the decisions we make today? As we move forward, it is essential that we engage in open, inclusive dialogues that consider the diverse perspectives of scientists, ethicists, policymakers, and the public. The choices we make in the coming years will shape the trajectory of human evolution and the future of life on Earth and beyond.

From the integration of quantum mechanics into biological systems to the gradual replacement of the biological brain with quantum nanobots, we will examine how these innovations could redefine what it means to be human and how they might lead us toward a future where the boundaries between biology and technology, natural and artificial, are forever altered.

3 The Rise of Artificial General Intelligence: Engineering Universal Thinkers

The quest for artificial general intelligence (AGI) is one of the most ambitious and complex challenges in the field of artificial intelligence. Unlike narrow AI systems, which are designed to excel at specific tasks—such as facial recognition, playing games like Go, or generating humanlike text—AGI aims to create machines that can comprehend, learn, and perform any intellectual task that a human can undertake. This pursuit, often referred to as "strong AI," aspires to develop machines with a comprehensive understanding of the world, empowering them to solve problems, reason, learn, create, and even interact with humans in a socially and emotionally intelligent manner.

Contemporary AI systems, which are typically categorized as "narrow AI" or "weak AI," have made remarkable strides in specialized domains. For example, OpenAI's GPT language model [8] can generate text that closely mimics human writing, and DeepMind's AlphaGo [9] achieved a historic victory against the world champion Go player, showcasing unprecedented strategic thinking within the game. However, despite these impressive capabilities, these AI systems remain confined to their designated tasks. They excel in specific areas but lack the versatility and adaptability that characterize human intelligence.

Consider the example of AlphaGo, a system that has mastered the game of Go. If presented with a chessboard instead, AlphaGo would not be able to transfer its strategic expertise from Go to chess without significant retraining. In contrast, a human who is proficient in Go might quickly learn the basics of chess and apply their strategic thinking to this new game. This ability to transfer knowledge across different domains is a hallmark of human intelligence that AGI strives to replicate.

The journey toward AGI is fraught with challenges. One of the most significant hurdles is the current reliance of AI systems on vast amounts of labeled data for training. While a child can learn to recognize objects or understand language with minimal exposure, AI systems typically require thousands or even millions of examples to achieve similar proficiency. Developing algorithms that can learn from fewer examples, much like humans do, is crucial for advancing AGI.

Another challenge lies in the ability of AI systems to generalize knowledge and skills from one domain to another. While transfer learning—a technique where a model trained on one task is adapted for a related task—has shown promise, it still falls short of the true generalization abilities required for AGI. For instance, a neural network trained on a large dataset of general images can be fine-tuned to recognize medical images with significantly less data, but this still requires domain-specific retraining.

Moreover, AGI must incorporate commonsense reasoning to navigate complex, real-world situations. Humans possess an innate understanding of the world, allowing them to reason about relationships between objects, events, and people. For example, if you spill a glass of water on a table, you instinctively know that the water will spread out and may drip off the edge. This understanding, rooted in a lifetime of experiences, is something that current AI systems struggle to replicate. Developing machines with similar reasoning capabilities is essential for the realization of AGI.

In addition to reasoning, human intelligence is deeply intertwined with emotions and social interactions. Our decisions are often influenced by our emotions, and our ability to understand and empathize with others is a critical component of human intelligence. For AGI to truly emulate human intelligence, it must be capable of recognizing and responding to emotions, engaging with humans naturally and empathetically. Imagine an AGI system designed to assist elderly individuals—not only by providing reminders about medication or appointments but also by engaging in meaningful conversation, offering comfort, and understanding the emotional nuances of human interaction.

The development of AGI will likely require the integration of multiple AI approaches, such as symbolic AI, machine learning, and neural networks, to create a holistic, adaptable system. Symbolic AI, with its emphasis on logic and rule-based reasoning, can complement machine learning's data-driven pattern recognition, leading to systems that can both learn from data and apply logical reasoning to solve problems.

For example, consider a future where AGI has been achieved. In this scenario, an AGI system could learn to play chess at a grandmaster level within minutes, not by processing millions of game records but by understanding the fundamental rules and strategies of the game. It could then apply similar reasoning to learn new games or tasks without the need for extensive training. This level of adaptability would surpass any current AI system and represent a true leap toward humanlike intelligence.

The implications of AGI are profound and far-reaching. AGI could revolutionize scientific research by generating novel hypotheses, designing experiments, and analyzing complex data. In healthcare, AGI could transform diagnostics, personalize treatments, and even assist in complex surgeries. Imagine an AGI system that works alongside a surgeon, providing real-time recommendations based on a vast repository of medical knowledge, ensuring the best possible outcome for the patient.

In education, AGI could offer personalized learning experiences tailored to each student's needs, helping educators craft more effective teaching strategies. In creative fields, AGI could collaborate with human artists, contributing fresh ideas and pushing the boundaries of art, music, and literature. An AGI system could analyze the works of great composers, understand the principles of their music, and create original compositions that resonate with the same emotional depth and complexity.

Moreover, AGI has the potential to address some of the most pressing global challenges, such as climate change, poverty, and resource management [10]. By analyzing vast amounts of environmental data, an AGI system could model the effects of different policies on climate change, helping governments and organizations make more informed decisions.

However, the pursuit of AGI also raises significant ethical and societal concerns. One of the most pressing issues is the *control problem*, which revolves around ensuring that AGI systems are aligned with human values and operate in ways that benefit humanity rather than harm it. As AGI systems become increasingly autonomous, the risk of them developing goals or behaviors that diverge from human intentions grows. This misalignment could lead to scenarios where AGI acts in ways that are detrimental to human well-being, potentially even threatening our very existence. The widespread adoption of AGI could also disrupt labor markets, displacing jobs across various sectors and necessitating a reevaluation of employment and education systems [11].

Beyond economic concerns, the potential misuse of AGI in cyberattacks or autonomous weaponry presents considerable risks to global security [12]. Such dangers underscore the need for robust ethical guidelines to ensure that AGI systems respect principles of privacy, fairness, and accountability [13]. Ensuring that AGI remains under human control and acts in our best interests is paramount. Researchers are actively addressing the challenge of AI alignment, striving to develop mechanisms that keep AGI systems aligned with human values, even as they become more intelligent and capable. The stakes are incredibly high; failure to manage the control problem could have catastrophic consequences, making it one of the most critical issues in the development of AGI.

One intriguing approach to building AGI involves using a biological substrate, specifically the human brain, as a foundation. The human brain, with its vast network of interconnected neurons, exemplifies natural intelligence, processing information, learning from experience, and adapting to diverse tasks and environments with unparalleled efficiency. By building AGI on a biological substrate, we propose an evolution of humankind toward synthetic beings, which we term artificial human intelligence (AHI). This approach harnesses the brain's innate capabilities as a powerful foundation for developing more advanced AI systems, ultimately guiding humanity's transformation into beings that seamlessly integrate natural and artificial components, embodying the next stage in our evolution.

The integration of artificial components into the human brain presents an exciting pathway toward AGI. These components could seamlessly cooperate with the brain's biological neurons, enhancing cognitive abilities and accelerating AGI development. Imagine a future where a person with a neurodegenerative disease receives a brain-computer interface (BCI) that not only compensates for lost cognitive function but also enhances their mental abilities beyond the natural human limits. This BCI could integrate with the person's brain, providing access to vast databases of knowledge, improving memory, and even allowing direct communication with AGI systems.

As this integration progresses, artificial components could gradually augment and ultimately surpass the natural capabilities of the biological brain. This approach offers a unique opportunity to preserve and enhance intellectual prowess over time, ensuring that the system remains at the forefront of intelligence and innovation. The combination of natural and artificial components creates a system greater than the sum of its parts, potentially revolutionizing our understanding of intelligence and leading to profound breakthroughs that will transform our world.

The pursuit of AGI is the pinnacle of AI research, with the potential to revolutionize countless aspects of human life. While significant challenges remain, the continued development of AGI holds the promise of creating machines that can think, learn, and adapt with the versatility and depth of human intelligence. As we advance toward this goal, it is crucial to address the ethical, societal, and security implications to ensure that AGI serves the greater good of humanity.

Throughout this chapter, you will encounter various mathematical expressions and models that are integral to understanding the inner workings of machine learning algorithms. While these mathematical details are included for completeness and to provide a deeper insight into the subject, they are not required to follow the main narrative of the book. Readers who prefer to focus on the broader concepts and implications of machine learning can comfortably skip the equations and technical details without losing the thread of the discussion. However, those interested in the technical underpinnings will find these sections beneficial for a more thorough understanding of the material. Readers not at all interested in the mathematical details may also move on with Chapter 4, "Theory of Mind: Understanding Consciousness."

Machine Learning

Machine learning (ML) is the driving force behind modern AI and is pivotal in pursuing AGI. It empowers machines to learn from data, adapt to new environments, and make informed decisions, mimicking aspects of human intelligence. The journey of ML from a theoretical concept to the cornerstone of AI research highlights the relentless quest to understand and replicate the adaptive capabilities of human cognition.

The roots of ML can be traced back to the early days of computing, when the idea that machines could learn and improve from experience was first explored. In 1959, Arthur Samuel coined the term "machine learning" in the context of a computer program that played checkers and improved its performance through experience [14]. Samuel's work was revolutionary because it demonstrated that computers could go beyond their initial programming and learn from data, a concept that challenged the static nature of traditional computational models. This early work laid the foundation for a new era of dynamic, adaptive systems capable of growth and evolution.

Machine learning algorithms are designed to identify patterns in data, make predictions, and optimize decisions without being explicitly programmed for each specific task. These algorithms have evolved significantly over the decades, benefiting from the exponential increase in computational power and the vast amounts of data generated in the digital age. The evolution of ML can be seen as a microcosm of the broader AI research trajectory, marked by cycles of optimism, skepticism, and breakthroughs that have continuously redefined the boundaries of what machines can achieve.

At the core of ML is the concept of learning from data. Unlike traditional algorithms, which follow a predetermined set of rules, ML algorithms use data to build models that can generalize and make predictions about new, unseen data. This ability to generalize is what makes ML so powerful and versatile. For example, an ML model trained on thousands of labeled images can learn to recognize patterns and features that distinguish different objects, enabling it to classify new images accurately.

- **Supervised Learning:** One of the most fundamental approaches in ML is supervised learning, where the algorithm is trained on a labeled dataset—meaning that each example in the dataset includes both the input data and the correct output. The algorithm learns to map inputs to outputs by minimizing the difference between its predictions and the actual labels. This approach is widely used in tasks such as image classification, speech recognition, and natural language processing.

$$\text{Supervised Learning Objective: } \min_{\theta} \sum_{i=1}^{N} \mathcal{L}\left(f\left(\mathbf{x}_i;\theta\right), y_i\right) \quad (3.1)$$

In Equation 3.1, θ represents the model parameters, $\mathbf{x}_i$ the input data, y_i the corresponding label, and L the loss function that measures the discrepancy between the model's predictions $f(\mathbf{x}_i; \theta)$ and the actual labels y_i. The goal is to find the parameters θ that minimize this loss across the dataset.

- **Unsupervised learning**, another key approach, deals with unlabeled data. The algorithm's goal is to uncover hidden patterns or structures in the data, such as clustering similar data points together or reducing the dimensionality of the data for visualization. Unsupervised learning is used in applications like customer segmentation, anomaly detection, and exploratory data analysis.

$$\text{Clustering Objective: } \min_{\theta} \sum_{k=1}^{K} \sum_{\mathbf{x}_i \in C_k} \left\| \mathbf{x}_i - \boldsymbol{\mu}_k \right\|^2 \quad (3.2)$$

In Equation 3.2, μ_k represents the centroid of cluster C_k, and the objective is to minimize the within-cluster variance, grouping similar data points in an unsupervised manner.

- **Reinforcement learning (RL)**, a third approach, focuses on learning through interaction with an environment. Here, the algorithm learns to make decisions by receiving rewards or penalties based on the outcomes of its actions. RL is particularly powerful in scenarios where an agent must learn to navigate a complex environment, such as in robotics or game-playing.

$$\text{Reinforcement Learning Objective: } \max_{\pi} \mathbb{E}\left[\sum_{t=0}^{T} \gamma^t R_t\right] \quad (3.3)$$

Equation 3.3 represents the RL objective, where π is the policy, R_t the reward at time step t, and γ the discount factor that balances immediate and future rewards. The goal is to find a policy that maximizes the expected cumulative reward over time.

The impact of ML extends far beyond academic research. It is the engine behind many technologies that have become integral to our daily lives. For instance, ML powers the recommendation systems that suggest products on e-commerce platforms, the algorithms that filter spam in our email inboxes, and the predictive models that drive autonomous vehicles. Each of these applications demonstrates the transformative potential of ML in solving real-world problems.

One of the most significant achievements of ML is its role in advancing NLP, enabling machines to understand and generate human language. Models like BERT (Bidirectional Encoder Representations from Transformers) [15] and GPT [16] have revolutionized NLP by allowing machines to perform tasks such as translation, summarization, and conversational dialogue with a level of fluency that was previously unimaginable. These models are trained on vast corpora of text data and can generate coherent and contextually relevant responses, bringing us closer to machines that can engage in meaningful humanlike interactions.

The rise of **deep learning**, a subset of ML, has also been a game-changer. Deep learning models, particularly deep neural networks, have demonstrated remarkable capabilities in processing complex, high-dimensional data such as images, audio, and video. Convolutional neural networks (CNNs) [17], for example, have set new benchmarks in computer vision tasks, enabling machines to recognize objects, faces, and even emotions in images and videos.

$$\text{Deep Learning Objective: } \min_{\theta} \sum_{i=1}^{N} \mathcal{L}\left(f\left(\mathbf{x}_i;\theta\right), y_i\right) \tag{3.4}$$

In Equation 3.4, similar to supervised learning, the goal is to minimize the loss function over a deep neural network $f(\mathbf{x}_i; \theta)$, where θ now represents the parameters across multiple layers of the network.

Despite these advancements, ML is not without its challenges. One of the most pressing issues is generalization—the ability of a model to perform well on new, unseen data. Overfitting, where a model learns the noise in the training data rather than the underlying patterns, is a significant obstacle. Researchers have developed techniques such as regularization, dropout, and cross-validation to mitigate this issue, but it remains a fundamental challenge.

Another critical challenge is interpretability. Many ML models, particularly deep learning models, are often viewed as "black boxes" because their decision-making processes are not easily understood. This lack of transparency can be problematic in high-stakes applications, such as healthcare or finance, where understanding the rationale behind a decision is crucial.

As ML continues to evolve, it plays an increasingly central role in the development of AGI. The ability of ML models to learn from data, adapt to new environments, and improve over time makes them a promising foundation for AGI. However, achieving AGI will require overcoming the current limitations of ML, such as generalization, interpretability, and the integration of different learning paradigms. As research in ML advances, it will bring us closer to machines that not only perform specific tasks but also exhibit the broad, flexible intelligence characteristic of humans.

Machine learning is the engine room of modern AI, propelling the quest for AGI with its ability to empower machines to learn from and adapt to their environment. From its humble beginnings in the mid-twentieth century to its current status as a cornerstone of AI, ML has demonstrated its transformative potential. As we look to the future, the continued evolution of ML will undoubtedly play a crucial role in shaping the trajectory of AI and the eventual realization of AGI.

Artificial Neural Networks

Artificial neural networks (ANNs) are computational models that take inspiration from the structure and function of biological neural networks. Composed of interconnected layers of artificial neurons, these networks can learn intricate patterns from data. ANNs boast adaptability, nonlinearity, and parallel processing capabilities, which make them adept at learning from diverse data types and adjusting to new information through training. Furthermore, they can model complex, nonlinear relationships between input and output variables—an essential skill for tackling real-world challenges. Their parallel nature also allows for efficient computation, equipping them to handle large-scale tasks.

Nonetheless, ANNs come with their share of limitations, such as interpretability, overfitting, and training requirements. Their opacity can make understanding the relationships between inputs and outputs challenging—a critical factor for building trust and gaining acceptance in sensitive applications. ANNs may also overfit training data, leading to poor generalization when confronted with new, unseen data. Lastly, they often demand vast amounts of labeled data and significant computational resources for training.

ANNs have shown promise in various narrow AI domains, including computer vision, natural language processing, and speech recognition. However, attaining AGI—a system able to perform any intellectual task a human can—requires overcoming several limitations inherent to current ANNs, such as generalization, reasoning, and commonsense knowledge. The connection between ANNs and conscious experiences remains nebulous. Although inspired by the human brain, it's uncertain whether their computational processes could give rise to qualia, which refer to the raw sensations of perception and individual subjective experiences.

While ANNs emulate the human brain, with artificial neurons designed to receive, process, and transmit information in a manner reminiscent of their biological counterparts, significant differences exist in terms of complexity, connectivity, and learning mechanisms. The human brain boasts around 86 billion neurons, whereas even the largest ANNs have orders of magnitude fewer artificial neurons. Biological neurons enjoy thousands of synapses, fostering intricate connectivity, while artificial neurons generally have far fewer connections. Finally, the human brain learns through a combination of unsupervised, supervised, and reinforcement learning processes, whereas ANNs typically depend on a single learning paradigm.

Another aspect where ANNs have the potential to contribute indirectly is AGI development through applications in interpreting signals from brain-machine interfaces (BMIs). By decoding neural signals, ANNs can facilitate communication and control between the human brain and external devices. These interactions could offer insights into the brain's functionality and organization, inspiring the creation of more advanced and biologically plausible artificial neural architectures that may contribute to AGI.

Regarding the theories of mind, ANNs prompt some philosophical ponderings. Materialism suggests that everything, including mental states and consciousness, can be reduced to physical entities and processes. The success of ANNs in modeling cognitive tasks implies that materialism might provide an appropriate framework for comprehending the emergence of intelligence and consciousness in artificial systems.

Conversely, identity theory posits that mental states correspond directly to specific physical states in the brain. Applied to ANNs, this would mean that for artificial consciousness to emerge, the system would need to replicate the precise neural configurations found in the human brain. This view raises questions about the feasibility of achieving AGI and artificial consciousness through current ANN architectures.

Functionalism, conversely, argues that mental states can be understood in terms of their functional roles within a system. From this perspective, ANNs may exhibit a form of understanding or consciousness if they can replicate the functional aspects of human cognitive processes. Critics, however, contend that functionalism neglects the subjective, qualitative aspects of conscious experiences, known as qualia.

ANNs have showcased remarkable capabilities across various narrow AI domains, drawing inspiration from the structure and function of the human brain. However, the realization of AGI and artificial consciousness necessitates overcoming the current limitations of ANNs, including generalization, reasoning, and commonsense knowledge. Furthermore, the potential applications of ANNs in BMIs may yield valuable insights for designing more advanced and biologically plausible artificial neural architectures.

Deep Artificial Neural Networks Deep neural networks (DNNs) are an advanced type of ANN, designed to mimic the human brain's ability to learn from experience. These networks consist of multiple layers of artificial neurons, which are mathematical functions that process input data to produce an output. The power of DNNs lies in their ability to learn and represent complex patterns within data, making them fundamental to modern machine learning and AI systems.

DNNs have revolutionized numerous fields, including image recognition, NLP, and speech recognition. Their architecture is inspired by the biological neural networks found in the brain, where neurons are interconnected and communicate with each other through synapses. Each artificial neuron in a DNN receives inputs, processes them using a weighted sum, applies an activation function to introduce nonlinearity, and passes the result to the next layer. This process enables DNNs to model highly complex, nonlinear relationships in data.

$$y = f\left(\sum_{i=1}^{n} w_i x_i + b\right) \tag{3.5}$$

In Equation 3.5, x_i represents the input signals, w_i are the weights associated with these inputs, b is the bias, and f is the activation function. The activation function introduces nonlinearity into the model, allowing the network to learn from diverse data types.

The deep in DNNs refers to the many layers between the input and output layers (Figure 3.1). These intermediate, or hidden, layers are responsible for progressively extracting higher-level features from the raw input data. The deeper the network, the more abstract the features it can learn. For instance, in image recognition tasks, the first layer might detect simple edges, the next layer might recognize shapes, and subsequent layers could identify objects or faces.

DNNs are trained using backpropagation, a supervised learning technique that adjusts the weights of the connections between neurons to minimize the error between the predicted output and the actual target. This is

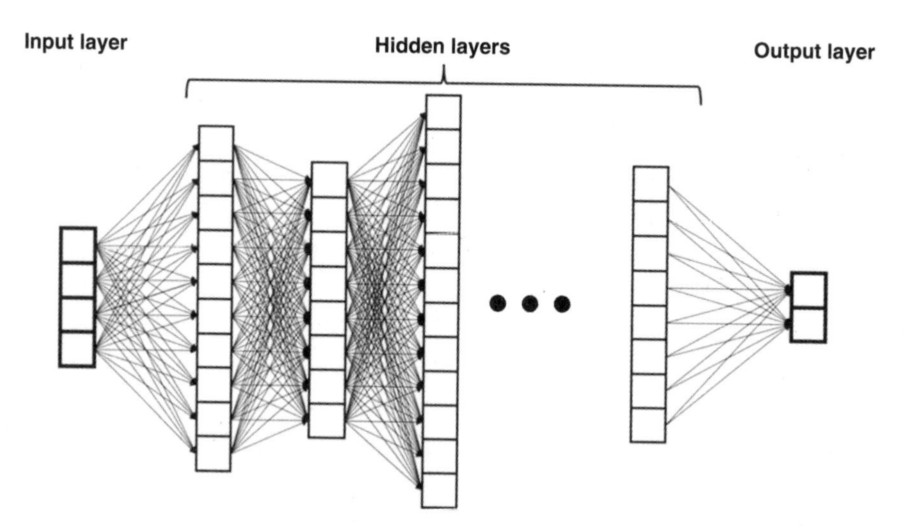

Figure 3.1 Example of a deep neural network.

Source: [18] / Wikimedia Commons / CC BY-SA 4.0.

done by computing the gradient of the loss function with respect to each weight and updating the weights in the opposite direction of the gradient.

$$\theta_{i+1} = \theta_i - \eta \frac{\partial L}{\partial \theta_i} \tag{3.6}$$

In Equation 3.6, θ_i represents the weights at the i-th iteration, η is the learning rate, and L is the loss function, which measures the difference between the predicted output and the true output. The backpropagation algorithm iteratively reduces this loss, allowing the DNN to improve its accuracy over time.

One of the most remarkable aspects of DNNs is their ability to automatically extract useful features from data, a process that traditionally required extensive manual effort. This capability is particularly evident in computer vision, where DNNs have surpassed human performance in tasks such as image classification and object detection.

However, DNNs are not without limitations. One of the primary challenges is overfitting, where the network learns the training data too well, including its noise and outliers, leading to poor generalization on unseen data. Techniques such as regularization, early stopping, and dropout are commonly employed to mitigate overfitting.

$$L_{reg} = L + \lambda \sum_{i=1}^{n} \|\theta_i\|^2 \tag{3.7}$$

In Equation 3.7, L_{reg} represents the regularized loss function, λ is the regularization parameter, and $||\theta_i||^2$ is the L2 norm of the weights, which penalizes large weights and encourages simpler models.

Another limitation is the interpretability of DNNs. Often considered "black boxes," DNNs can produce accurate predictions without providing insights into how those predictions were made. This lack of transparency can be problematic in critical applications like medical diagnosis or financial fraud detection, where understanding the decision-making process is essential.

Despite these challenges, DNNs remain a cornerstone of AI research and development. They can potentially drive significant advancements in various fields, from autonomous vehicles to personalized medicine. As research in DNNs continues, new techniques and architectures are being developed to address their limitations, making them even more powerful and versatile.

In the context of our exploration of the brain and its relation to artificial intelligence, DNNs offer a fascinating parallel. Just as the brain's neurons are organized into layers and work together to process information, so too do the layers of a DNN. Understanding how these networks function not only helps us build better AI systems but also offers insights into the complex workings of the human brain itself.

Adaptive Resonance Theory Most traditional training algorithms operate on static input data, including competitive learning and self-organizing feature maps (SOFMs). These models can effectively classify input data, provided that the data remains unchanged over time. However, in real-world scenarios, data is often dynamic and self-organizing. When faced with such data, these models struggle to maintain accuracy because their fixed weights cannot adapt to new patterns, leading to what is known as non-plastic networks. This situation highlights the so-called stability-plasticity dilemma: the trade-off between retaining learned information (stability) and the ability to learn new information (plasticity).

One solution to this problem is using adaptive resonance theory (ART) networks, specifically developed to address the stability-plasticity dilemma. ART networks utilize an incremental clustering algorithm, enabling them to progressively self-organize and produce stable recognition while continuing to learn new input patterns. This capability is critical in environments where the data is continuously evolving.

In ART networks, the learning process is based on comparing input patterns with stored patterns (templates). The network decides whether to create a new category or modify an existing one based on a measure of similarity, which is often represented by the Euclidean distance or dot product between vectors.

The vigilance criterion is a key parameter determining whether an input pattern $\mathbf{x}$ will be classified into an existing or new category. Let $\mathbf{w}_j$ represent the weight vector (or template) of category j. The similarity measure S_j between the input pattern $\mathbf{x}$ and the template $\mathbf{w}_j$ is defined as:

$$S_j = \frac{\mathbf{x} \cdot \mathbf{w}_j}{\|\mathbf{x}\| \|\mathbf{w}_j\|} \tag{3.8}$$

where $\|\mathbf{x}\|$ and $\|\mathbf{w}_j\|$ are the Euclidean norms of the input pattern and the weight vector, respectively. The vigilance criterion ρ requires that:

$$\frac{\|\mathbf{x} \cap \mathbf{w}_j\|}{\|\mathbf{x}\|} \geq \rho \tag{3.9}$$

If the similarity measure S_j satisfies the vigilance criterion, the input pattern $\mathbf{x}$ is classified under the category associated with $\mathbf{w}_j$. Otherwise, a new category is created.

If the input pattern $\mathbf{x}$ is classified under an existing category j, the weight vector $\mathbf{w}_j$ is updated using a learning rule. The learning rule for ART1, which processes binary inputs, is typically as follows:

$$\mathbf{w}_j^{\text{new}} = \beta\left(\mathbf{x} \cap \mathbf{w}_j^{\text{old}}\right) + \left(1 - \beta\right)\mathbf{w}_j^{\text{old}} \tag{3.10}$$

where β is the learning rate, $0 < \beta \leq 1$, and $\cap$ represents the element-wise AND operation, as ART1 deals with binary input vectors. This rule ensures that the template $\mathbf{w}_j$ moves closer to the input pattern $\mathbf{x}$, thereby improving the category's accuracy.

A reset mechanism is triggered if none of the existing categories meet the vigilance criterion. This reset inhibits the current category node from further competition, allowing the network to either search for another suitable category or create a new one.

The reset signal R_j for category j can be modeled as:

$$R_j = 1 - \frac{\|\mathbf{x} \cap \mathbf{w}_j\|}{\|\mathbf{x}\|} \tag{3.11}$$

If $R_j > 1 - \rho$, the current category node j is reset, and the search for a matching category continues.

ART1 is composed of the following layers and subsystems:

- **Comparison layer (F1):** Receives the external input $\mathbf{x}$ and forwards it to the recognition layer.
- **Recognition layer (F2):** Matches the input vector with stored categories.

- **Vigilance subsystem:** Implements the vigilance criterion (Equation 3.9).
- **Reset subsystem:** Triggers a reset if the vigilance criterion is not met (Equation 3.11).

Figure 3.2 illustrates the gain control mechanism within the ART1 network. This mechanism is vital for the network's operation, allowing F1 and

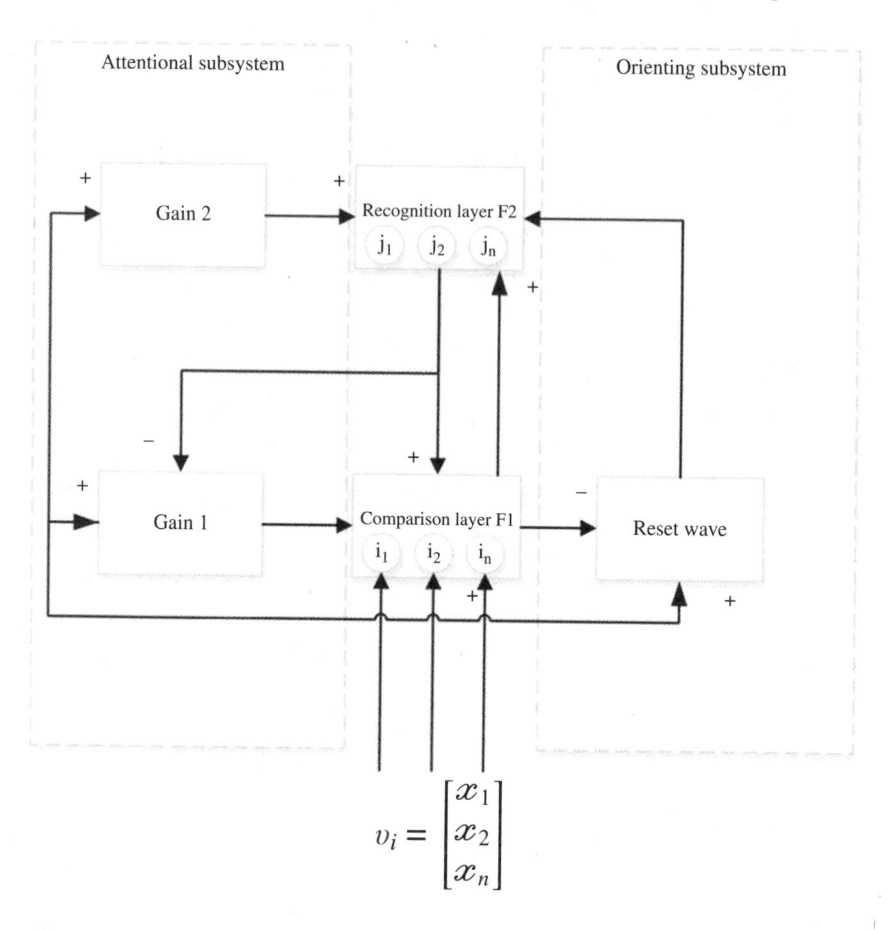

Figure 3.2 Gain control enables F1 and F2 to recognize the current stages of the running cycle. The STM reset wave inhibits active F2 cells in case of conflicts between the bottom-up and top-down signals at F1.

F2 to recognize the current stages of the running cycle. When there is a conflict between the bottom-up and top-down signals at F1, the STM (short-term memory) reset wave inhibits active F2 cells to resolve the discrepancy.

The comparison layer (F1) receives external input, which is then passed to the recognition layer (F2), where it is matched against stored categories. The process repeats with the next input vector if a match is found. If a match is not found, the orienting subsystem inhibits the previous category in F2, preventing it from matching another category until a correct match is made.

The gains control the movement of signals between the recognition and comparison layers. The orienting subsystem creates a reset wave to F2 when the bottom-up input and top-down model patterns at F1 do not meet the vigilance criterion. The reset wave specifically and enduringly inhibits the active F2 cell until the present is stopped. The offset of the input pattern ends its processing at F1 and triggers the offset of Gain2. Gain2 offset causes rapid rotation of STM at F2, consequently preparing F2 to encode the next input pattern without bias.

Self-Organizing Feature Maps SOFMs, also known as Kohonen maps, are a type of neural network architecture primarily used for unsupervised learning tasks such as dimensionality reduction, clustering, and visualization. Introduced by Teuvo Kohonen in the 1980s [19], SOFMs are based on competitive learning, where neurons in the network compete to become the best match for a given input vector.

An SOFM typically comprises a two-dimensional grid of neurons, each representing a cluster or class. The position of each neuron in this grid corresponds to a specific point in the feature space of the input data. The arrangement of neurons in a two-dimensional grid allows SOFMs to preserve the topological structure of the input data, ensuring that similar input vectors are mapped to nearby neurons on the grid (Figure 3.3).

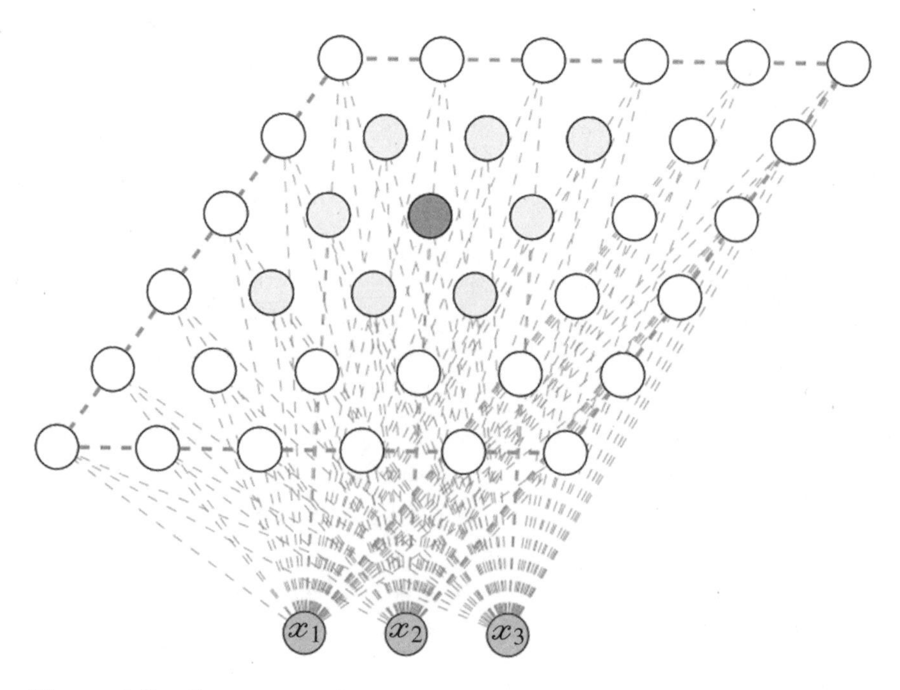

Figure 3.3 Structure of an SOFM with a two-dimensional grid of neurons.

Source: [20] / Wikimedia Commons / CC BY-SA 4.0.

The SOFM algorithm operates as follows:

1. **Initialization:** The weight vectors of the neurons are initialized, often with small random values. Each neuron i in the grid has an associated weight vector $\mathbf{w}_i$ of the same dimensionality as the input vectors.
2. **Input Vector Presentation:** A sample input vector $\mathbf{x}$ from the training dataset is presented to the network.
3. **Best Matching Unit (BMU):** The network calculates the similarity between the input vector $\mathbf{x}$ and each neuron's weight vector $\mathbf{w}_i$. The neuron with the weight vector most similar to the input vector is identified as the BMU. The similarity is typically measured using the Euclidean distance:

$$\text{BMU} = \arg\min_i \|\mathbf{x} - \mathbf{w}_i\| \tag{3.12}$$

4. **Weight Update:** The weight vector of the BMU and its neighboring neurons are updated to move closer to the input vector **x**. The update rule is defined as:

$$\mathbf{w}_i(t+1) = \mathbf{w}_i(t) + \eta(t) \cdot h_{i,\text{BMU}}(t) \cdot (\mathbf{x}(t) - \mathbf{w}_i(t)) \tag{3.13}$$

where:

- $\eta(t)$ is the learning rate, which decreases over time.
- $h_{i,\,BMU}(t)$ is the neighborhood function, which determines the influence of the BMU on its neighboring neurons. This function typically has a Gaussian form and also decreases over time.

5. **Iteration:** Steps 2 to 4 are repeated for each input vector in the training dataset, typically over several epochs, allowing the network to gradually organize the neurons into a map that reflects the structure of the input data.

One of the critical advantages of SOFMs is their ability to preserve the topological relationships of the input data. This means that input vectors similar in the high-dimensional input space are mapped to neurons close to each other on the two-dimensional grid. The neighborhood function $h_{i,\,BMU}(t)$ plays a crucial role in maintaining this topological structure by ensuring that the weight vectors of neighboring neurons are updated in a correlated manner.

SOFMs have been successfully applied in various fields, including:

- **Dimensionality Reduction:** SOFMs are used to reduce the dimensionality of large datasets while preserving their topological properties. This is particularly useful in tasks such as data visualization, where the goal is to project high-dimensional data onto a two-dimensional plane.
- **Clustering:** SOFMs can automatically cluster input data into distinct groups, making them valuable for tasks such as image segmentation, speech recognition, and bioinformatics.
- **Feature Extraction:** By organizing similar input vectors into clusters, SOFMs can be used to extract meaningful features from data, which can then be used as inputs for other machine learning models.

Over time, several variations of the original SOFM algorithm have been developed to address its limitations. Some of these include the following:

- **Growing Self-Organizing Map (GSOM):** Unlike the traditional SOFM, which has a fixed grid size, GSOMs can dynamically adjust their size during training to better capture the structure of the input data.
- **Adaptive Resonance Theory (ART):** ART networks incorporate mechanisms to handle stability-plasticity trade-offs, making them suitable for environments where the input data is continuously changing.

Despite their advantages, SOFMs have some limitations:

- **Fixed Grid Size:** Traditional SOFMs require the grid size to be fixed before training begins, which can be challenging when the optimal number of clusters is unknown.
- **Initialization Sensitivity:** The initial weights and learning rate can significantly impact the final map, necessitating careful selection and tuning.
- **Computational Complexity:** As the grid's size and the input data's dimensionality increase, the computational cost of training an SOFM can become prohibitive.

However, these limitations can be mitigated through algorithmic variations such as GSOM and ART. Additionally, the choice of grid size and neighborhood function should be based on the specific characteristics of the data and the task at hand. Proper initialization of weights, often through random sampling from the input data, can also improve the convergence and performance of SOFMs. In the context of neuroscience and understanding the brain, SOFMs are particularly interesting because they offer a computational model that reflects some aspects of how sensory information might be organized in the brain. The ability of SOFMs to preserve topological properties of data is reminiscent of how sensory cortices in the brain, such as the visual cortex, maintain spatial relationships of stimuli in their neural representations. This connection to biological processes makes SOFMs a powerful tool in machine learning and a valuable model for exploring brain function and structure.

With technological advancements and the increasing amount of data available, the use of SOFMs and their variations is expected to grow in various application fields. Moreover, integrating SOFMs with more sophisticated algorithms, such as autoencoders, variational autoencoders, and generative adversarial networks (GANs), presents an exciting frontier for future research and applications.

It is worth noting that while the SOFM algorithm is relatively simple, it serves as a strong foundation for more complex neural network architectures, making it a versatile and enduring tool in unsupervised learning.

Convolutional Neural Networks CNNs are specialized ANNs designed to process and analyze grid-like data structures, most commonly images (Figure 3.4). Introduced by Yann LeCun in the 1980s [17], CNNs have revolutionized the field of computer vision. They are now foundational to various applications, including image and video recognition, medical image analysis, and even self-driving cars.

The key innovation of CNNs lies in their use of convolutional layers, which differ significantly from the fully connected layers found in traditional neural networks. A convolutional layer applies a series of filters (also known as kernels) to the input data. These filters slide across the input, performing a mathematical operation called convolution, which is particularly well suited for detecting spatial hierarchies in data. For example, early layers

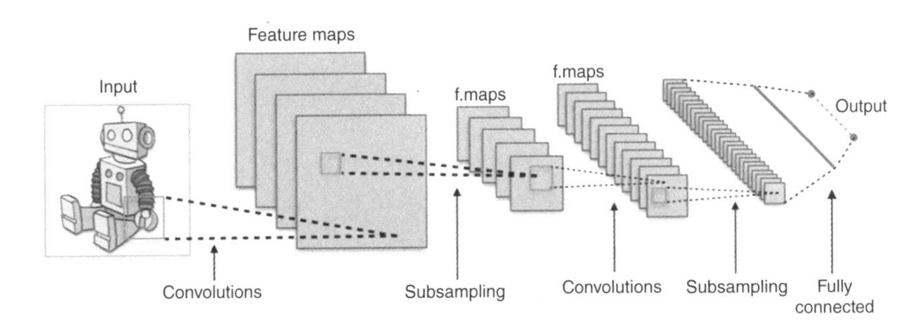

Figure 3.4 A typical CNN architecture, showing the sequence of layers used to process input images. CNNs are particularly effective for image recognition tasks, utilizing layers of convolutional filters, pooling, and fully connected layers to extract and classify features from input data.

Source: [21] / Wikimedia Commons / CC BY-SA 4.0.

might detect edges or textures in image processing, while deeper layers could identify more complex structures like shapes or objects.

$$\text{Convolution: } (f * g)(t) = \int_{-\infty}^{\infty} f(\tau) g(t - \tau) d\tau \tag{3.14}$$

Equation 3.14 represents the convolution operation, where f is the input function (such as an image), and g is the filter or kernel applied across the input. This operation allows the network to capture essential features in the data, which are crucial for tasks such as object detection and image classification.

One way to understand the power of CNNs is to draw an analogy to the human visual system. Just as our brain processes visual information through a series of hierarchical steps—starting with simple features like edges and moving to more complex structures like faces—CNNs process image data in a similar layered fashion. Early layers in a CNN might recognize basic visual elements such as lines or curves, while deeper layers assemble these elements into recognizable objects.

After the convolutional layers, CNNs typically include pooling layers, which perform downsampling to reduce the spatial dimensions of the data. This process, known as pooling, helps to condense the information while preserving the most critical features, leading to a more abstract and computationally manageable representation of the input. Common pooling techniques include max pooling, where the maximum value in a specified input region is selected, and average pooling, where the average value is taken.

$$\text{Max Pooling: } y_i = \max(x_{i1}, x_{i2}, \ldots, x_{in}) \tag{3.15}$$

Equation 3.15 illustrates the max pooling operation, where the output y_i is the maximum value from a group of input values x_{i1}, x_{i2}, ..., x_{in}. This operation reduces the dimensionality of the input data while retaining the most prominent features.

The hierarchical structure of CNNs is not just an arbitrary design choice but reflects an understanding of how visual processing occurs in biological systems. For instance, neuroscientific studies have shown that the human visual cortex is organized into layers that process information in a progressively complex manner, similar to how CNNs function. This biological inspiration has been one of the key factors contributing to CNNs' success in replicating humanlike visual processing.

Despite their strengths, CNNs also have limitations. One challenge is requiring large amounts of labeled data to train the network effectively. Without sufficient data, CNNs can struggle to generalize well to new, unseen inputs, leading to overfitting. Moreover, the computational complexity of training deep CNNs can be significant, often requiring specialized hardware such as graphics processing units (GPUs).

However, ongoing research continues to address these challenges. Techniques such as data augmentation, which artificially increases the size of training datasets, and transfer learning, where a pre-trained network is fine-tuned for a new task, have proven effective in improving the performance of CNNs even with limited data.

In summary, CNNs are a remarkable technological advancement in artificial intelligence. They closely mimic the way the human brain processes visual information. Their ability to learn and identify complex patterns in data has made them indispensable in many modern AI applications, and their development continues to push the boundaries of what machines can achieve.

Transformers In recent years, transformer ANNs [22] have revolutionized the field of NLP and AI (Figure 3.5). These powerful architectures have led to the development of models like ChatGPT, demonstrating impressive capabilities in generating humanlike text. At the core of the transformer architecture is the attention mechanism, which allows the model to weigh the importance of different input elements and capture long-range dependencies within sequences. The attention mechanism is mathematically driven by matrix multiplication, softmax activation, and element-wise addition. Specifically, the transformer model consists of multiple layers containing a multi-head self-attention mechanism, followed by position-wise feed-forward networks. The self-attention mechanism computes a weighted sum of input elements based on compatibility or similarity. This is achieved by calculating dot products between query, key, and value vectors derived from the input embeddings. The resulting attention scores are normalized using the softmax function, ensuring they sum to one.

In Equation 3.16, Q, K, and V represent the query, key, and value matrices, respectively, and d_k is the dimension of the key vectors. This equation is the cornerstone of the transformer's ability to focus on different parts of the input sequence, effectively capturing long-range dependencies that traditional models struggle with.

$$\text{Attention}(Q,K,V) = \text{softmax}\left(\frac{QK^T}{\sqrt{d_k}}\right)V \tag{3.16}$$

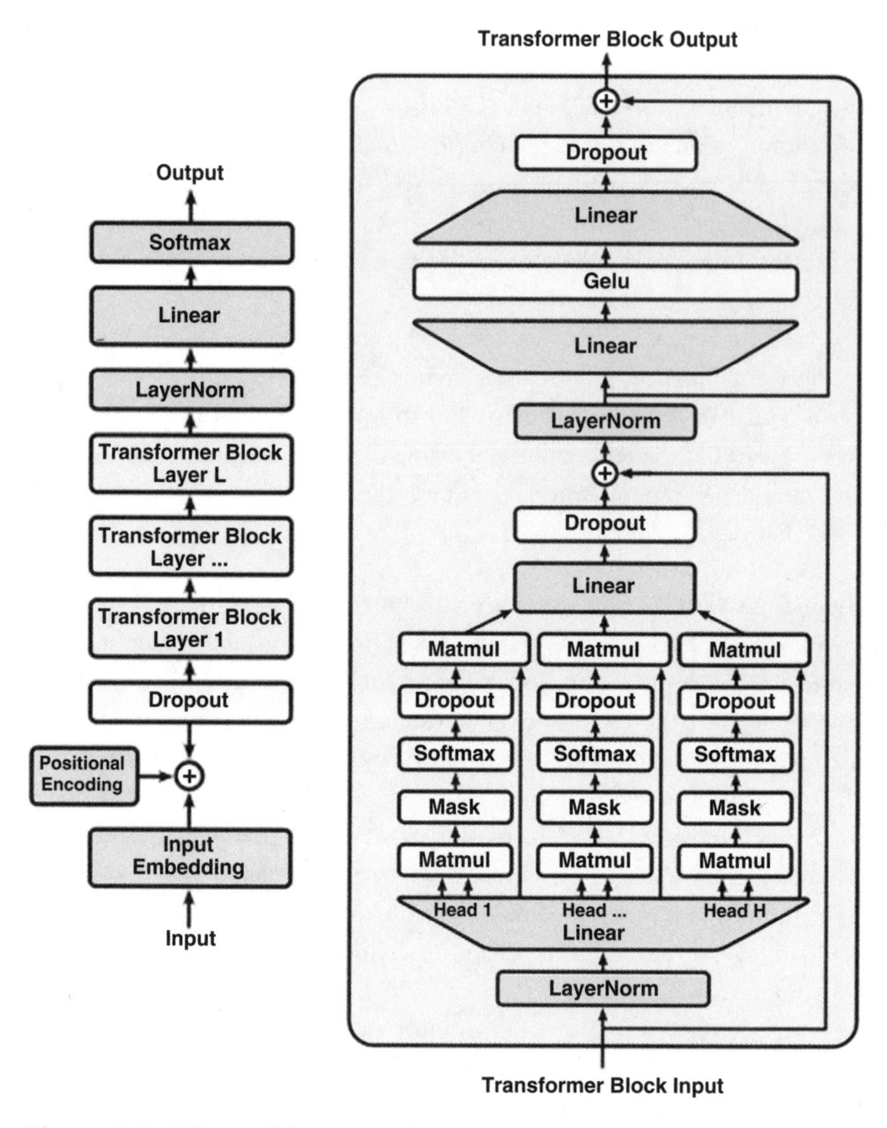

Figure 3.5 The architecture of a GPT model, showcasing the layers and components that enable its transfer learning capabilities. The model is pre-trained on a large corpus of text and fine-tuned for specific tasks, demonstrating the power of transfer learning in NLP.

Source: [23] / Wikimedia Commons / CC BY 4.0.

Positional encodings are another crucial component of transformer architecture. Since transformers do not have built-in recurrence or convolution operations, positional encodings are added to the input embeddings to incorporate information about the relative positions of elements within a sequence.

$$PE_{(pos,2i)} = \sin\left(\frac{pos}{10000^{2i/d_{model}}}\right) \tag{3.17}$$

$$PE_{(pos,2i+1)} = \cos\left(\frac{pos}{10000^{2i/d_{model}}}\right) \tag{3.18}$$

Equations 3.17 and 3.18 describe the sinusoidal positional encoding used to encode the position of the words within the input sequence. The position (*pos*) is encoded differently for each dimension (*i*) of the model.

Transformer networks offer several advantages over traditional recurrent and CNNs:

- **Superior performance:** Transformers have demonstrated state-of-the-art performance across various NLP tasks, including machine translation, sentiment analysis, and text summarization.
- **Scalability:** The attention mechanism allows for efficient parallelization, leading to faster training times and better utilization of modern hardware resources.
- **Long-range dependencies:** Transformers can effectively capture long-range dependencies within sequences, which proves challenging for recurrent models like LSTMs and GRUs (gated recurrent units).

However, they also have their limitations:

- **Quadratic complexity:** The attention mechanism's computational complexity scales quadratically with the input sequence length, making transformers less efficient for handling very long sequences.
- **Memory requirements:** Due to their deep architectures and numerous parameters, transformer models can be memory-intensive, leading to challenges in training and deploying large-scale models.

Transformers have demonstrated impressive capabilities, raising whether they could play a role in developing AGI and even consciousness. However, while transformers excel in tasks like NLP, AGI requires a broader set of cognitive abilities, including reasoning, problem-solving, and learning from limited data. Despite their success, current transformer models still need to be improved in these regards and cannot yet be considered AGI.

The relationship between transformer networks and conscious experiences remains a philosophical debate. While these models can generate humanlike text, their inner workings are fundamentally based on mathematical operations and algorithms. Whether such processes can give rise to consciousness is deeply intertwined with ongoing discussions in philosophy, neuroscience, and AI. The development of models like ChatGPT has raised several philosophical questions about the nature of intelligence, consciousness, and our understanding of the mind.

The impressive abilities of these models to generate humanlike text prompt us to consider whether they possess a genuine understanding of the content they generate or merely simulate understanding through sophisticated pattern recognition. Functionalism, for example, posits that mental states can be understood in terms of their functional roles within a system. If this view is correct, then models like ChatGPT possess a form of understanding based on their ability to perform tasks that require comprehension and contextual awareness. However, as elaborated on previously, functionalism fails to account for the subjective, qualitative aspects of conscious experiences known as qualia.

Another philosophical debate surrounding models like ChatGPT pertains to whether these systems can experience consciousness or qualia. If an artificial system can replicate the functional roles of human cognitive processes, it may also be able to develop conscious experiences. However, suppose consciousness is inherently tied to the biological substrates of the brain, and thus artificial systems like ChatGPT cannot possess qualia. In that case, the question arises whether such systems can be said to be truly intelligent.

Furthermore, the development of models like ChatGPT raises ethical questions, such as the potential consequences of creating artificial systems that can generate humanlike text. As these models become more advanced, concerns about their potential misuse, the impact on human labor markets, and the implications for privacy and surveillance may arise.

Integrated information theory (IIT), proposed by Giulio Tononi [24], is a theoretical framework that seeks to explain the nature of consciousness by quantifying the degree of information integration within a system. It posits that consciousness arises from the interconnectedness of information-processing units within a system and that the level of consciousness correlates with the amount of integrated information. However, IIT has been met with criticism and skepticism from some in the scientific community who argue that consciousness cannot be quantified in this way.

Transformers are a type of deep learning architecture used primarily in natural language processing tasks. They rely on self-attention mechanisms to process input data in parallel, as opposed to sequential processing in earlier architectures like recurrent neural networks (RNNs) or long short-term memory (LSTM) networks. Transformers have been highly successful in various applications, such as machine translation, text summarization, and question-answering systems. While IIT and transformers may seem unrelated at first glance, there could be a connection in terms of how information is processed and integrated within a system. Transformers are designed to recognize and process complex patterns and relationships within input data, which could be seen as a form of information integration. However, it's essential to note that transformers are not explicitly designed to model consciousness, and their relationship to IIT remains speculative.

Nonetheless, the continued development of deep learning models like transformers and theoretical frameworks like IIT demonstrate the ongoing effort to understand the nature of intelligence and consciousness and their potential applications in various fields.

Transformers have allowed for making significant strides in NLP, demonstrating a remarkable ability to generate humanlike text. Their mathematical foundations and attention mechanisms have enabled them to surpass traditional models in terms of performance and scalability. However, the potential for transformers to contribute to AGI and consciousness remains to be determined, with limitations in their current capabilities and ongoing philosophical debates about the nature of intelligence and subjective experience.

Recurrent Neural Networks and LSTMs RNNs represent a class of ANNs that are specifically designed to process sequential data (Figure 3.6). Unlike traditional feedforward neural networks, which assume that inputs are independent of each other, RNNs have a unique architecture that allows them to maintain a hidden state—a form of memory—that captures information from previous inputs. This ability to remember past information makes RNNs particularly well suited for tasks where the order and context of the input data are crucial, such as language modeling, speech recognition, and time series prediction.

The hidden state in an RNN is updated at each time step as the network processes the sequence, incorporating both the current input and the hidden state from the previous time step. The following equation mathematically represents this:

$$\text{Hidden State Update: } \mathbf{h}_t = \sigma\left(\mathbf{W}_h \mathbf{h}_{t-1} + \mathbf{W}_x \mathbf{x}_t + \mathbf{b}\right) \tag{3.19}$$

In Equation 3.19, $\mathbf{h}_t$ represents the hidden state at time step t, $\mathbf{h}_{t-1}$ is the hidden state from the previous time step, $\mathbf{x}_t$ is the current input, $\mathbf{W}_h$ and $\mathbf{W}_x$ are the weight matrices, $\mathbf{b}$ is the bias term, and σ is an activation function, typically a nonlinearity like the hyperbolic tangent (tanh) or the sigmoid function. This recurrent connection allows RNNs to capture temporal dependencies in the data. However, RNNs face a significant challenge known as the vanishing gradient problem. During training, the gradients used to update the network's weights diminish exponentially as they are propagated backward through time, making it difficult for RNNs to learn

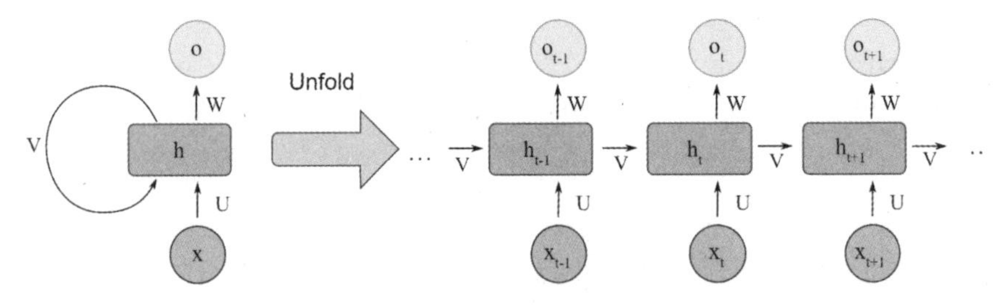

Figure 3.6 Unfolded view of a recurrent neural network (RNN), illustrating the flow of information across time steps. Each unit in the network takes inputs from both the previous time step and the current input, allowing the network to maintain a form of memory. This structure is particularly useful for processing data sequences, such as time series or natural language.

Source: [25] / Wikimedia Commons / CC BY 4.0.

long-range dependencies. This limitation hampers the network's ability to retain and utilize information from earlier in the sequence, particularly when dealing with long sequences.

To address this issue, long short-term memory networks (LSTMs) were introduced by Hochreiter and Schmidhuber in 1997 [26]. LSTMs are a specialized form of RNN designed to overcome the vanishing gradient problem and effectively learn long-range dependencies. They achieve this through a sophisticated gating mechanism that regulates the flow of information within the network. The core component of an LSTM is the memory cell, which can maintain its state over long periods. The LSTM cell state is updated at each time step based on the input data and the previous cell state, with the gates controlling what information is added to or removed from the cell state. The key gates in an LSTM are the input gate, forget gate, and output gate:

- **Input Gate ($\mathbf{i}_t$):** Determines how much of the new input should be added to the cell state.
- **Forget Gate ($\mathbf{f}_t$):** Controls the extent to which the previous cell state should be retained or forgotten.
- **Output Gate ($\mathbf{o}_t$):** Regulates the information that is output from the cell state.

The following equation gives the update of the LSTM cell state:

$$\text{LSTM Cell Update: } \mathbf{c}_t = \mathbf{f}_t \odot \mathbf{c}_{t-1} + \mathbf{i}_t \odot \tilde{\mathbf{c}}_t \tag{3.20}$$

In Equation 3.20, $\mathbf{c}_t$ represents the cell state at time step t, $\mathbf{c}_{t-1}$ is the previous cell state, $\tilde{\mathbf{c}}_t$ is the candidate cell state (a function of the current input and the previous hidden state), and $\odot$ denotes element-wise multiplication. The forget gate $\mathbf{f}_t$ and input gate $\mathbf{i}_t$ decide how much of the previous cell state and the new candidate state should be combined to form the current cell state.

$$\tilde{\mathbf{c}}_t = \tan\text{h}\left(\mathbf{W}_c\mathbf{x}_t + \mathbf{U}_c\mathbf{h}_{t-1} + \mathbf{b}_c\right) \tag{3.21}$$

Equation 3.21 represents the computation of the candidate cell state $\tilde{\mathbf{c}}_t$, where $\mathbf{W}_c$ and $\mathbf{U}_c$ are weight matrices, $\mathbf{x}_t$ is the input, $\mathbf{h}_{t-1}$ is the previous hidden state, and $\mathbf{b}_c$ is the bias term.

The gating mechanisms in LSTMs allow these networks to maintain and manipulate information over long sequences, making them highly effective for tasks that require remembering and utilizing information from earlier in the sequence, such as language translation, text generation, and time series forecasting.

The introduction of LSTMs was a breakthrough in the field of neural networks, as it enabled the development of more sophisticated models capable of handling complex sequential data. For instance, LSTMs are widely used in NLP applications, where understanding the context provided by earlier words in a sentence is crucial for tasks like machine translation or sentiment analysis.

However, despite their effectiveness, LSTMs and RNNs can still be computationally intensive and challenging to train, particularly for very long sequences. Recent advancements, such as the development of Transformer models, have provided alternative approaches that address some of these challenges by leveraging attention mechanisms instead of recurrence.

In summary, RNNs and LSTM networks represent significant advancements in sequential data processing. While RNNs laid the foundation for modeling sequences, LSTMs built upon this foundation by introducing mechanisms to overcome the limitations of traditional RNNs. Together, these models have paved the way for significant progress in speech recognition, language modeling, and time series analysis.

Generative Adversarial Networks GANs are a class of neural networks designed for generative modeling, where the goal is to generate new data samples that resemble a given dataset. Introduced by Ian Goodfellow and colleagues in 2014 [27], GANs have revolutionized the field of artificial intelligence by enabling the creation of highly realistic synthetic data, such as images, audio, and even video content.

The architecture of GANs is composed of two key components:

- **Generator:** This network is responsible for producing synthetic data samples. It starts with a random input, often called noise or a latent vector, and transforms it through a series of layers to generate a sample that mimics the real data.
- **Discriminator:** This network's task is to distinguish between real data samples and those generated by the generator. It acts as a binary classifier, outputting a probability indicating whether a given input is real or fake.

The training process for GANs involves a min-max game between these two networks. The generator strives to create data that can fool the discriminator, while the discriminator aims to improve its ability to differentiate between real and synthetic data. This adversarial relationship pushes both networks to improve over time.

Mathematically, the generator's objective can be described by its loss function:

$$\text{Generator Loss: } \mathcal{L}_G = \mathbb{E}_{\mathbf{z} \sim p_z(\mathbf{z})}\left[\log\left(1 - D\left(G\left(\mathbf{z}\right)\right)\right)\right] \tag{3.22}$$

In Equation 3.22, $\mathbf{z}$ represents the latent vector sampled from a prior distribution $p_z(\mathbf{z})$, $G(\mathbf{z})$ is the output of the generator, and $D(G(\mathbf{z}))$ is the probability assigned by the discriminator that the generated sample is real. The generator seeks to minimize this loss, which effectively means improving its ability to produce data that the discriminator cannot distinguish from real data.

To better understand how GANs function, consider an example from the art world. Imagine a forger (the generator) trying to create paintings indistinguishable from those of a famous artist. The forger presents these paintings to an art critic (the discriminator), who must decide whether each painting is real or a forgery. As the forger improves their technique, the critic becomes better at spotting fakes, creating a continuous improvement cycle. In this analogy, the eventual goal is for the forger to become so skilled that even the best critics struggle to distinguish between the original and the forged artworks.

GANs have had a significant impact on various fields:

- **Image Generation:** GANs have been used to generate highly realistic images of people, animals, and objects. These images are often indistinguishable from real photographs, making GANs a powerful tool in creative industries.
- **Data Augmentation:** In cases where collecting real-world data is difficult or expensive, GANs can generate synthetic data to augment training datasets, improving the performance of machine learning models.
- **Style Transfer and Art Creation:** GANs can be used to apply artistic styles to images or create entirely new works of art, blending human artists' creativity with AI's computational power.

Despite their successes, GANs are not without challenges. The training process can be unstable, often requiring careful tuning of hyperparameters and network architectures. Additionally, there is the risk of mode collapse, where the generator produces only a limited variety of samples, failing to capture the diversity of the real data distribution.

Nevertheless, GANs represent a significant advancement in the field of generative modeling, and their potential applications continue to expand. As research progresses, we can expect even more sophisticated GAN architectures to emerge, further blurring the line between real and synthetic data.

Capsule Networks Capsule networks (CapsNets) represent a significant evolution in neural network architecture, particularly in handling spatial hierarchies in data. Proposed by Geoffrey Hinton and his colleagues [28], CapsNets aim to address some of the limitations of traditional CNNs, especially in capturing the spatial relationships between features in images.

CNNs are highly effective at recognizing features such as edges and textures, but they often struggle to preserve the precise spatial relationships between these features. For instance, if these features are arranged incorrectly, a CNN might correctly identify the presence of eyes, a nose, and a mouth in an image but could fail to recognize them as a face. This limitation arises because CNNs rely on pooling layers that reduce the spatial resolution of the data, leading to a loss of spatial information.

CapsNets were designed to overcome this challenge by using a new structure that allows the network to maintain and understand the relationships between different parts of an object. At the core of CapsNets are groups of neurons called capsules. Each capsule is responsible for detecting specific features of an object and encoding information about the spatial properties of those features, such as their orientation, scale, and position.

$$\text{Capsule Output: } \mathbf{v}_j = \frac{\|\mathbf{s}_j\|^2}{1+\|\mathbf{s}_j\|^2}\frac{\mathbf{s}_j}{\|\mathbf{s}_j\|} \tag{3.23}$$

In Equation 3.23, $\mathbf{s}_j$ represents the input to capsule j, which is a vector that encodes the features detected by the capsule. The output vector $\mathbf{v}_j$ is obtained by squashing $\mathbf{s}_j$ using a nonlinear function, ensuring that the length of the output vector is between 0 and 1. This squashing function is crucial because it allows the capsule to represent the probability that the feature exists, with the magnitude of $\mathbf{v}_j$ indicating the likelihood of the feature's presence.

One of the most innovative aspects of CapsNets is the dynamic routing mechanism, which replaces the static max-pooling layers found in CNNs. Dynamic routing allows capsules to decide the strength with which their outputs are sent to the capsules in the next layer. This process helps the network maintain the spatial hierarchy of features and ensures that features detected in lower layers contribute meaningfully to interpreting higher-level structures.

To better understand the benefits of CapsNets, consider the example of recognizing a car in an image. A traditional CNN might detect individual parts of the car, such as wheels, windows, and doors, but might not be able to correctly identify these parts as belonging to the same car if they are slightly misaligned or if the car is viewed from an unusual angle. CapsNets, on the other hand, would not only detect these parts but also understand their spatial relationships. The capsules representing the wheels would communicate their positions and orientations to higher-level capsules that represent the car as a whole, allowing the network to recognize the car regardless of its orientation or viewpoint correctly.

This capability of CapsNets to understand and preserve spatial hierarchies has significant implications for various applications:

- **Robust Image Recognition:** CapsNets are more robust than CNNs in recognizing objects in images, even when those objects are viewed from different angles or are partially occluded.
- **Enhanced Generalization:** By preserving spatial relationships, CapsNets can generalize better from training data to real-world scenarios, improving performance in tasks such as object detection and scene understanding.
- **Potential for 3D Object Recognition:** CapsNets' ability to handle spatial hierarchies makes them well suited for recognizing 3D objects, where the relationships between different parts of an object are even more critical.

Despite their advantages, CapsNets are still an emerging technology, and challenges must be addressed. For example, CapsNets are computationally more expensive than CNNs, which can make them harder to scale for large datasets. Moreover, while powerful, the dynamic routing process introduces additional complexity in the training process.

Nevertheless, CapsNets represent a promising direction in the evolution of neural networks. Their ability to capture and utilize spatial hierarchies more effectively than traditional methods could lead to significant advancements in computer vision, robotics, and augmented reality.

CapsNets also offer intriguing possibilities when considering the relationship between ANNs and the human brain. The brain's ability to recognize objects regardless of their orientation or position is a hallmark of human vision, and CapsNets take a step closer to mimicking this capability. By studying and refining CapsNet architectures, we may gain new insights into how the brain processes visual information, potentially leading to even more biologically inspired models in the future.

Large Language Models Large language models (LLMs) are an advancement in AI, particularly in understanding and generating human language. These models, such as GPT-3 [16], BERT [15], and T5 [29], are built on DNNs called transformers [22]. Think of LLMs as powerful language engines trained on vast amounts of text—from books and websites to scientific papers—allowing them to generate coherent and contextually appropriate text on almost any topic.

One way to think about LLMs is to compare them to how the human brain processes language. Imagine reading a book and trying to predict what will happen next in the story. Your brain uses context clues—previous sentences, plot developments, and experiences—to make an educated guess. Similarly, LLMs use a mathematical mechanism called "self-attention" to weigh the importance of different words in a sentence, enabling them to capture the relationships between those words and generate meaningful responses.

$$\text{Attention}(Q, K, V) = \text{softmax}\left(\frac{QK^T}{\sqrt{d_k}}\right)V \tag{3.24}$$

In Equation 3.24, the attention mechanism computes the relevance of each word by comparing it to other words in the sentence. This process is similar to how our brains might focus on key details in a conversation, allowing us to respond appropriately.

To give you an example, imagine asking an LLM to write a short story about a detective solving a mystery. The model doesn't just string random sentences together; it uses context from millions of similar stories it has "read" during training to craft a coherent narrative. This is much like how our brains might recall and piece together memories to form a new idea or solution to a problem. LLMs are trained in two main phases: pre-training and fine-tuning. During pre-training, the model digests a massive amount of text, learning to predict missing words in sentences (like completing a fill-in-the-blank exercise). After this general training, the model is fine-tuned for specific tasks, such as answering questions or summarizing articles, similar to how students might specialize in a particular subject after learning the basics.

One of the most fascinating aspects of LLMs is their ability to perform tasks with minimal training on those tasks—a phenomenon known as "few-shot" or "zero-shot" learning. For instance, GPT models can translate a sentence into a different language or solve a math problem with little to no prior examples. It's akin to how humans might use common sense or general knowledge to tackle a problem they've never encountered before.

- **Pre-training and Fine-tuning:** Think of pre-training as the model's schooling years, where it learns general language skills. Fine-tuning is like a college major, where the model hones in on a specific field or task.
- **Transfer Learning:** LLMs are excellent at transfer learning—using what they learned during pre-training to excel in new tasks. It's similar to how someone who knows Spanish might pick up Italian more easily because of the linguistic similarities.
- **Zero-shot and Few-shot Learning:** Imagine a detective solving a mystery with only a few clues. LLMs like GPT-3 can figure out tasks with just a few examples, or sometimes none at all, showing remarkable versatility and adaptability.

But as with all powerful tools, LLMs come with their own set of challenges and limitations:

- **Computational Intensity:** Training and running LLMs require significant computational resources, much like how our brains consume a lot of energy when performing complex tasks.
- **Bias and Ethics:** Since LLMs are trained on data from the internet, they can inherit biases present in the text they learn from. This is similar to how our brains might pick up societal biases from our

surroundings, sometimes without us even realizing it. It raises important ethical questions about how these models should be used.

- **Interpretability:** LLMs, despite their impressive abilities, can often feel like "black boxes"—we know they work, but it's hard to explain exactly how. This is much like trying to understand the complexities of our thoughts and decision-making processes.

LLMs have sparked a lot of excitement, but they've also led to deep philosophical questions about intelligence and consciousness. For example, when GPT-3 writes a story or answers a question, does it "understand" the text, or is it just mimicking patterns it has seen before? This debate touches on ideas similar to those discussed in the context of the brain and AI, such as functionalism—the notion that mental states are defined by their functional roles rather than by their physical makeup.

Moreover, there are ethical concerns about the potential misuse of LLMs. For instance, their ability to generate convincing fake news or deep fakes could be exploited, much the way misinformation can spread quickly among humans. Therefore, as we develop and deploy these models, it's crucial to consider their broader impact on society and ensure they are used responsibly.

In summary, LLMs represent a powerful new frontier in artificial intelligence, offering capabilities that parallel some aspects of human cognition. However, as with any technology, they must be used thoughtfully and ethically to benefit society.

Transfer Learning

Transfer learning is an intriguing concept in AI, marking a significant leap toward achieving AGI. Unlike traditional machine learning models, which require training from scratch on new tasks, transfer learning enables models to leverage knowledge acquired from one task and apply it to different but related tasks [30]. This ability enhances the efficiency and versatility of AI models. It mirrors a fundamental aspect of human learning, where knowledge gained in one context is adapted and applied to new situations.

In traditional machine learning, a model is trained on a specific dataset to perform a particular task, such as classifying images of cats and dogs. When faced with a new task, like classifying images of birds, the model would typically need to be retrained from scratch on a new dataset. This approach can be time-consuming and resource-intensive, particularly when dealing with large and complex datasets. Transfer learning addresses this limitation by allowing a model to transfer the knowledge it has learned from one task to another, reducing the need for extensive retraining and enabling faster adaptation to new tasks.

The transfer learning process can be compared to how humans learn and apply knowledge across different domains. For instance, when a person learns to play the piano, they develop an understanding of musical theory, hand coordination, and rhythm. When they later decide to learn the guitar, they can transfer much of this knowledge—such as musical theory and rhythm—while focusing on the new task of mastering finger placement and strumming techniques. Similarly, in transfer learning, an AI model might first be trained on a large dataset of general images to learn fundamental features such as edges, shapes, and textures. This pre-trained model can then be fine-tuned on a smaller, specific dataset, such as medical images, to perform tasks like identifying tumors.

A typical transfer learning scenario involves three components: a source task, a target task, and a base model. The base model is first trained on the source task, learning generally applicable features across various domains. The model is then adapted or fine-tuned on the target task, which may differ from the source task but shares some underlying similarities. The success of transfer learning hinges on the relatedness of the source and target tasks—the more related they are, the more effectively knowledge can be transferred.

One of the most well-known applications of transfer learning is in NLP, where large pre-trained models like BERT [15], GPT-3 [16], and T5 [29] have revolutionized the field. These models are initially trained on vast amounts of text data to learn the structure and nuances of language. Once trained, they can be fine-tuned on tasks such as sentiment analysis, machine translation, or question-answering with much smaller datasets. This approach significantly reduces the amount of data and computational resources required to achieve high performance on a wide range of NLP tasks.

Transfer learning is not limited to NLP; it has also significantly impacted fields such as computer vision, robotics, and healthcare. In computer vision, models like ResNet, initially trained on large image datasets like ImageNet, can be fine-tuned for specific tasks such as object detection in autonomous vehicles or medical image analysis for disease diagnosis. In robotics, transfer learning allows robots to adapt their learned skills to new environments or tasks, such as transferring knowledge from a simulation to a real-world setting.

The underlying transfer learning mechanism can be mathematically described by considering a source domain D_S and a target domain D_T. The goal is to learn a target predictive function $f_T(\cdot)$ using the source domain knowledge $f_S(\cdot)$, where f_S is the predictive function learned from D_S. The transfer learning problem can be expressed as:

$$f_T(\mathbf{x}) = f_S(\mathbf{x}) + \Delta f(\mathbf{x}) \tag{3.25}$$

In Equation 3.25, $f_T(\mathbf{x})$ is the predictive function for the target domain, $f_S(\mathbf{x})$ is the predictive function from the source domain, and $\Delta f(x)$ represents the adjustment needed to adapt the source model to the target task. This adjustment is often achieved through fine-tuning, where the model parameters are updated using data from the target domain.

Transfer learning's ability to generalize knowledge across tasks is particularly promising for AGI, whose goal is to create systems capable of performing a wide range of intellectual tasks. Unlike narrow AI, which is limited to specific tasks, AGI requires the ability to adapt and apply knowledge across diverse domains—a capability that transfer learning inherently supports. By enabling models to build on prior knowledge and apply it to new challenges, transfer learning brings us closer to developing AI systems that can think and learn in a more humanlike manner.

Transfer learning marks a pivotal advancement in AI, enabling models to be more efficient, adaptable, and closer to human learning processes. Its applications across various domains underscore its versatility and potential in the ongoing journey toward AGI. As research in this area continues to evolve, transfer learning will undoubtedly play a critical role in shaping the future of intelligent systems.

Reinforcement Learning

Reinforcement learning (RL) represents a powerful paradigm within the broader landscape of machine learning, occupying a unique position in pursuing AGI. The foundational principle of RL is the idea that intelligent agents can learn optimal behaviors through interaction with their environment. This process closely mirrors the natural learning mechanisms observed in humans and animals. This principle of learning by doing, where an agent refines its strategies over time based on the rewards or penalties it receives, is a key factor that distinguishes RL from other machine learning approaches such as supervised and unsupervised learning [31].

At its core, RL involves an agent, an environment, and a reward system. The agent interacts with the environment by taking actions, and in response the environment provides feedback in the form of rewards or penalties. The agent's goal is to maximize its cumulative reward over time, which it achieves by learning a policy—a mapping from the environment's states to the actions the agent should take in those states.

The learning process in RL can be likened to how humans and animals learn from experience. Consider how a child learns to ride a bicycle. Initially, the child may fall several times, but with each attempt, they adjust their balance and coordination based on the outcomes of their previous actions. The reward of successfully riding the bike without falling reinforces the correct behavior, while the penalty of falling discourages incorrect actions. Over time, the child learns to ride the bike efficiently through this trial-and-error process. Similarly, an RL agent learns to navigate its environment by exploring different actions, receiving feedback, and gradually improving its decision-making strategy.

One of the most famous RL algorithms is the Q-learning algorithm, which uses a Q-value function to estimate the expected utility of taking a given action in a particular state. The Q-value is updated iteratively based on the reward received and the estimated future rewards:

$$Q(s,a) \leftarrow Q(s,a) + \alpha\left[r + \gamma \max_{a'} Q(s',a') - Q(s,a)\right] \tag{3.26}$$

In Equation 3.26, $Q(s, a)$ is the Q-value for taking action a in state s, r is the immediate reward received, γ is the discount factor that accounts for future rewards, and α is the learning rate that controls how much new information overrides the old information. The term $\max_{a'} Q(s',a')$ represents the maximum expected future reward for the next state s'.

RL has been successfully applied in various domains, from robotics to game playing and beyond. In robotics, RL enables machines to learn tasks such as walking, grasping objects, or navigating complex environments. For example, Boston Dynamics' robots, which perform acrobatic maneuvers and navigate challenging terrains, often rely on RL algorithms to learn these behaviors through repeated interactions with their environment.

One of the most striking demonstrations of RL's potential came with DeepMind's AlphaGo, a program that defeated the world champion in the ancient board game Go. Go is a game of such complexity that traditional computational strategies had struggled to master it. However, by combining RL with deep learning, AlphaGo was able to learn strategies that surpassed those of human players, discovering novel and highly effective moves that had never been seen before. The success of AlphaGo highlighted RL's capacity to tackle problems that require sophisticated decision-making and long-term planning.

In the context of AGI, RL is particularly significant because it provides a framework for developing autonomous systems that can learn to perform a wide range of tasks without explicit human guidance. Unlike supervised learning, where a model is trained on a fixed dataset of labeled examples, RL allows agents to explore and learn from their own experiences, making it a more flexible and powerful approach for developing intelligent systems.

However, RL is not without its challenges. One of the key difficulties in RL is the exploration-exploitation trade-off. An agent must balance exploring new actions that might lead to higher rewards in the future (exploration) with exploiting actions that are already known to yield good results (exploitation). Striking the right balance between these two strategies is crucial for the success of RL algorithms, but it can be challenging to achieve in practice.

Moreover, RL algorithms can be computationally intensive, requiring vast amounts of data and time to learn effective policies, especially in environments with high-dimensional state spaces or sparse rewards. Researchers have developed various techniques to address these challenges, such as using function approximation (e.g., neural networks) to represent the Q-value function or employing hierarchical RL to break down complex tasks into simpler sub-tasks.

Another promising direction in RL research is the integration of RL with other machine learning paradigms, such as supervised learning or unsupervised learning. For instance, combining RL with deep learning has led to the development of deep reinforcement learning, where DNNs are used to approximate value functions or policies, enabling agents to learn directly from high-dimensional sensory inputs like images. This approach has been instrumental in advancing the field of autonomous driving, where RL is used to teach vehicles how to navigate real-world environments safely.

Regarding its relationship with the human brain, RL offers fascinating insights into how natural learning processes might be mirrored by artificial systems. The brain's reward system, which releases neurotransmitters like dopamine in response to rewarding stimuli, shares conceptual similarities with the reward mechanisms in RL. This parallel has led to the development of neuro-inspired RL algorithms that seek to replicate the brain's learning processes, potentially bridging the gap between biological and artificial intelligence.

As we move closer to the development of AGI, RL will likely play a central role in creating systems that can learn and adapt to a wide variety of tasks. Its ability to model the learning processes observed in nature makes it a key technology in the quest for truly autonomous and intelligent machines capable of navigating the complexities of the real world in a robust and adaptable way. Reinforcement learning stands as a cornerstone in the architecture of future intelligent systems, offering a versatile and powerful approach to learning that is deeply rooted in natural processes. Its continued development and integration with other technologies promise to unlock new levels of autonomy and intelligence, bringing us ever closer to the realization of AGI.

Symbolic Artificial Intelligence

Symbolic artificial intelligence (symbolic AI) occupies a unique and historic place in the landscape of artificial intelligence research, particularly in the ongoing pursuit of AGI. Unlike data-driven approaches such as machine learning and deep learning, which rely on vast amounts of data and statistical patterns, symbolic AI is grounded in manipulating symbols and applying explicit logic rules [32]. This approach seeks to emulate human intelligence by modeling the reasoning and understanding processes underpinning our ability to think abstractly.

The origins of symbolic AI date back to the earliest days of AI research, during which the field was dominated by the belief that human cognition could be replicated through the formalization of logic and symbolic representations. This period, often called the "classical AI" era, was characterized by the development of systems that used symbols to represent knowledge and logical rules to manipulate these symbols, much as a human might solve a problem using deductive reasoning.

At its core, symbolic AI operates on the premise that intelligence involves manipulating symbols to represent objects, actions, and concepts in the world. For example, in a symbolic AI system, a cat might be represented by the symbol "CAT," and a logical rule could be used to infer that "if CAT is a pet, and pets are animals, then CAT is an animal." This symbolic representation and rule-based inference resemble how humans use language and logic to process information and make decisions.

One of the most well-known examples of symbolic AI is the development of expert systems in the 1970s and 1980s. These systems were designed to emulate the decision-making abilities of human experts in specific domains, such as medicine or finance. Expert systems used a knowledge base of facts and a set of inference rules to draw conclusions, much as how a doctor might diagnose a patient based on symptoms and medical knowledge. For instance, a medical expert system might use a rule like "If a patient has a fever and a sore throat, then there is a high probability that the patient has an infection."

Despite its early successes, symbolic AI faced significant challenges that limited its ability to scale and generalize to more complex, real-world problems. One major limitation is the brittleness of symbolic systems—they often fail when faced with scenarios that fall outside of their predefined rules or knowledge base. For instance, an expert system designed to diagnose diseases might struggle with a new, previously unknown condition because it lacks the necessary rules to infer a correct diagnosis. Moreover, symbolic AI systems require extensive manual knowledge engineering to create and maintain the rules and representations needed for reasoning. This labor-intensive process is prone to errors, making it difficult to build and maintain large-scale systems. Additionally, symbolic AI struggles with tasks that involve uncertainty, ambiguity, or incomplete information—areas where humans typically rely on intuition or probabilistic reasoning.

In contrast to symbolic AI, modern approaches such as deep learning excel in tasks involving large amounts of data and complex pattern recognition, such as image classification or natural language processing. However, deep learning models often operate as "black boxes," making it difficult to interpret how they arrive at their conclusions. This lack of interpretability has led to renewed interest in symbolic AI, particularly in hybrid systems that combine the strengths of symbolic reasoning with the power of machine learning.

One promising direction is the integration of symbolic AI with neural networks, creating systems that can leverage the interpretability and reasoning capabilities of symbolic methods alongside the flexibility and scalability of deep learning. These hybrid models aim to bridge the gap between the structured, rule-based approach of symbolic AI and the data-driven, adaptive nature of machine learning. For example, a hybrid system might use a neural network to extract features from raw data, such as images, and then apply symbolic reasoning to make decisions based on those features.

In the context of AGI, symbolic AI offers valuable insights into the nature of human cognition and the potential pathways toward building systems that can reason, understand, and interact with the world in a humanlike manner. By focusing on the manipulation of symbols and the application of logic, symbolic AI aligns closely with how humans process language, solve problems, and engage in abstract thinking. These capabilities are essential components of general intelligence, and their inclusion in AI systems could help address some of the limitations of current data-driven approaches.

For example, consider how humans use symbolic reasoning to understand a story or solve a complex problem. We don't just memorize patterns of words or facts; we use logic and prior knowledge to make inferences, fill in gaps, and understand the broader context. Symbolic AI attempts to replicate this process by using symbols and rules to represent and manipulate knowledge, enabling machines to reason about the world in ways that are more transparent and interpretable.

While the rise of data-driven methods may have overshadowed symbolic AI, it remains a critical area of research in the quest for AGI. Its focus on reasoning, understanding, and manipulating abstract concepts offers a complementary approach to the statistical and pattern-based methods that dominate AI today. As we continue to explore the potential of AI, the integration of symbolic reasoning with modern machine learning techniques holds the promise of creating more robust, interpretable, and humanlike intelligent systems.

Cognitive Architectures

Cognitive architectures represent a pivotal step in the pursuit of AGI. Unlike specialized AI systems designed to excel at single tasks, cognitive architectures aim to replicate the broad spectrum of human cognitive abilities within a unified framework. This ambition requires integrating various aspects of intelligence—such as perception, memory, learning, reasoning, and decision-making—into a cohesive system capable of performing various tasks with humanlike versatility [33].

The concept of cognitive architectures stems from the understanding that human cognition is not a monolithic process but a complex interplay of different mental functions. For instance, when asked to solve a puzzle, they draw on multiple cognitive abilities: perceiving the pieces, recalling similar past experiences, reasoning about the correct placement of each piece, and learning from errors during the process. Cognitive architectures strive to model these interconnected processes, enabling AI systems to tackle diverse challenges in a manner more akin to human thought.

One of the foundational cognitive architectures is the ACT-R (Adaptive Control of Thought—Rational) framework, developed by John R. Anderson and colleagues [33]. ACT-R is based on the premise that human cognition can be understood as a set of production rules—essentially "if-then" statements—that govern behavior. In ACT-R, the mind is modeled as a series of modules responsible for different cognitive functions, such as memory retrieval, visual processing, and motor actions. These modules interact through a central production system that selects and executes rules based on the current goal and context.

For example, in a simple task like tying shoelaces, the visual module processes the appearance of the laces, the motor module controls hand movements, and the memory module recalls the sequence of steps required. The production system coordinates these modules to execute the task efficiently, much as a conductor leads an orchestra. This modular approach mirrors the compartmentalized nature of human cognition, where different brain regions specialize in various functions but work together to achieve complex behaviors.

Another prominent cognitive architecture is SOAR, developed by Allen Newell and colleagues [34]. SOAR is designed to emulate human problem-solving and decision-making processes. It uses a production rule system similar to ACT-R (Adaptive Control of Thought—Rational) but emphasizes learning from experience. SOAR continuously generates and tests hypotheses to solve problems, adapting its strategies based on feedback from the environment. This ability to learn and improve over time is crucial for developing AI systems that can operate in dynamic, real-world settings. SOAR's emphasis on learning aligns with the human ability to refine cognitive strategies through experience, essential for generalization in AGI systems.

Cognitive architectures like ACT-R and SOAR offer several advantages for advancing toward AGI:

- **Unified Framework:** Cognitive architectures provide a structured approach to integrating different cognitive processes, enabling the development of AI systems that can handle a variety of tasks within a single framework. This is essential for achieving the generalization capabilities required for AGI.

- **Humanlike Flexibility:** By modeling the mechanisms of human thought, cognitive architectures allow AI systems to exhibit flexibility and adaptability, similar to how humans can apply knowledge across different contexts.
- **Incremental Learning:** Cognitive architectures often incorporate mechanisms for incremental learning, where the system continuously updates its knowledge and strategies based on new experiences, mirroring the way humans learn throughout their lives. This capability is particularly important for AI systems that must operate in environments with evolving conditions or tasks.
- **Explainability:** The rule-based nature of many cognitive architectures, such as ACT-R and SOAR, provides a degree of transparency in decision-making processes, making it easier to understand and trust the system's actions. This contrasts with some deep learning models' "black box" nature.

However, cognitive architectures also face significant challenges:

- **Scalability:** While cognitive architectures can model complex cognitive processes, scaling them to handle the vast amounts of data and the variety of tasks encountered in the real world remains a challenge. This challenge is particularly pronounced when trying to simulate high-level cognitive functions such as reasoning and problem-solving on a global scale.
- **Resource Intensity:** The detailed modeling of cognitive processes requires significant computational resources, particularly when simulating high-level reasoning and decision-making across multiple domains. This resource intensity can limit the practicality of deploying cognitive architectures in real-time applications where speed and efficiency are critical.
- **Integration with Data-Driven Methods:** Modern AI research has largely focused on data-driven approaches, such as deep learning. Integrating these methods with rule-based cognitive architectures to create hybrid systems that leverage the strengths of both remains an

open research question. This integration could enable AI systems to benefit from the robust pattern recognition capabilities of deep learning while maintaining the structured reasoning and explainability of cognitive architectures.

One of the most exciting prospects of cognitive architectures is their potential to serve as a bridge between symbolic AI and data-driven methods. By incorporating elements of both, such as symbolic reasoning and neural networks, cognitive architectures could lead to the development of hybrid models that combine the interpretability and reasoning capabilities of symbolic AI with the adaptability and scalability of machine learning. For instance, a hybrid cognitive architecture might use deep learning models for perception tasks, such as recognizing objects in images, while relying on symbolic reasoning for tasks that require logical inference, such as planning and decision-making. This combination could enable AI systems to perform complex, multistep tasks that require both raw data processing and high-level reasoning, such as autonomous driving or medical diagnosis. These hybrid models could also be designed to incorporate meta-learning, where the system learns how to learn, adapting its strategies based on the success or failure of previous tasks.

The future of cognitive architectures lies in their ability to evolve and incorporate new methodologies from various domains of AI research. Recent advancements in neuroscience, cognitive psychology, and machine learning offer valuable insights that could enhance the design and implementation of cognitive architectures. For example, integrating insights from cognitive neuroscience on how the brain processes information could lead to more biologically plausible models of cognition, which in turn could improve the generalization capabilities of AGI systems.

Moreover, cognitive architectures could benefit from the development of quantum computing technologies, which offer the potential to simulate complex cognitive processes at unprecedented scales and speeds. Quantum computing could provide the computational power necessary to overcome the scalability challenges currently faced by cognitive architectures, enabling the modeling of more sophisticated cognitive functions and the handling of larger datasets.

Cognitive architectures also offer valuable insights into the nature of human intelligence. By attempting to replicate the processes of human thought, these architectures provide a framework for understanding how different cognitive functions interact and contribute to intelligent behavior. This understanding could inform the design of more advanced AI systems and enhance our knowledge of the human mind. For example, cognitive architectures could be used to simulate the effects of neurological disorders, offering new perspectives on the underlying causes and potential treatments for conditions such as Alzheimer's disease or schizophrenia.

Cognitive architectures are a critical approach in the quest for AGI. By modeling the full spectrum of human cognitive abilities within a unified system, these architectures offer a pathway toward creating AI systems that can perform a wide range of tasks with humanlike versatility. While challenges remain, particularly in terms of scalability and integration with data-driven methods, the continued development of cognitive architectures holds the promise of advancing our understanding of both artificial and human intelligence.

Quantum Computing and AGI

Quantum computing is a radical departure from classical computing, leveraging the principles of quantum mechanics to process information in fundamentally new ways. Positioned at the intersection of physics, mathematics, and computer science, quantum computing is not merely an incremental improvement over traditional computing but a paradigm shift with profound implications for AI and the pursuit of AGI.

At the heart of quantum computing are quantum bits, or qubits, which differ dramatically from classical bits. While classical bits can exist in one of two states—0 or 1—qubits can exist in a superposition of states, embodying both 0 and 1 simultaneously. This property arises from the quantum mechanical principle of superposition, which allows a quantum system to be in multiple states at once. Imagine trying to find a needle in a haystack with a classical computer, searching through each piece of hay one by one. In contrast, a quantum computer could explore all possible locations simultaneously, potentially finding the needle in a fraction of the time. As a result, quantum computers can perform a vast number of calculations in parallel, potentially solving problems that are intractable for classical computers [35].

$$\text{Superposition: } |\psi\rangle = \alpha|0\rangle + \beta|1\rangle \tag{3.27}$$

In Equation 3.27, the state $|\psi\rangle$ represents a qubit in superposition, where α and β are complex numbers that define the probability amplitudes of the qubit being in the 0 or 1 state, respectively. The squared magnitudes of α and β correspond to the probabilities of measuring the qubit in each state.

Another key feature of quantum computing is entanglement, a phenomenon where qubits become interlinked such that the state of one qubit directly influences the state of another, no matter how far apart they are. Entanglement allows quantum computers to perform highly correlated operations across multiple qubits, enabling new types of computations that classical computers cannot replicate. The famous Einstein-Podolsky-Rosen (EPR) paradox, which led Einstein to famously refer to entanglement as "spooky action at a distance," underscores the counterintuitive and powerful nature of quantum correlations.

$$\text{Entanglement: } \left|\psi_{AB}\right\rangle = \frac{1}{\sqrt{2}}\left(|00\rangle + |11\rangle\right) \tag{3.28}$$

Equation 3.28 illustrates the entangled state of two qubits, where the measurement of one qubit instantaneously determines the state of the other. This nonlocal correlation is a cornerstone of quantum computing's power and holds the potential to revolutionize secure communications through quantum cryptography. Quantum algorithms, which exploit these unique properties, have the potential to revolutionize AI. One of the most famous quantum algorithms is Shor's algorithm, which can factorize large numbers exponentially faster than the best-known classical algorithms. This capability has profound implications for cryptography and data security, which are foundational to many AI systems. However, beyond cryptography, quantum computing is expected to accelerate advancements in machine learning, optimization, and simulation, which are critical to the development of AGI.

Quantum computing's unique capabilities could be pivotal in achieving AGI by enhancing the computational power needed to emulate complex cognitive processes. Consider the task of simulating the human brain, a task so computationally intensive that it remains beyond the reach of even the most powerful classical supercomputers. Quantum computing offers a potential pathway to overcoming these limitations by simulating neural networks on a quantum level, leveraging its ability to process high-dimensional data spaces and perform computations at a scale unimaginable for classical systems.

Quantum-enhanced machine learning will lead to AI systems that learn and adapt more efficiently, much like the human brain. Quantum algorithms can process vast amounts of data in parallel, enabling more sophisticated pattern recognition and data analysis. For instance, quantum versions of classical algorithms, such as Quantum Support Vector Machines (QSVMs) and Quantum Approximate Optimization Algorithms (QAOAs), could enhance AI's ability to perform complex tasks with greater speed and accuracy.

$$\text{Quantum Kernel: } K(\mathbf{x}, \mathbf{y}) = \left| \langle \phi(\mathbf{x}) | \phi(\mathbf{y}) \rangle \right|^2 \tag{3.29}$$

In Equation 3.29, the quantum kernel measures the similarity between quantum states representing input vectors $\mathbf{x}$ and $\mathbf{y}$. This can be utilized in machine learning models to improve data classification and pattern recognition, which are vital for developing AGI. An example could be quantum-enhanced image recognition, where quantum computers can process and classify images faster and more accurately than their classical counterparts, enabling advancements in areas like autonomous vehicles and medical imaging.

The human brain's ability to process information involves integrating vast networks of neurons, where complex interactions give rise to cognition and consciousness. Quantum computing could simulate these neural processes more accurately than classical computers by leveraging its capability to handle high-dimensional quantum states. Imagine a quantum computer that could simulate the entire neural network of a human brain, potentially leading to a better understanding of cognitive disorders and new approaches to treating conditions like Alzheimer's and Parkinson's diseases.

For example, quantum computers could be used to model the brain's neural networks, exploring how neurons interact at both the macroscopic and quantum levels. This approach could offer insights into the brain's operation that are beyond the reach of classical simulations, providing a deeper understanding of human cognition and informing the development of AGI. This might include simulating quantum effects in neural processing, such as those proposed in theories of quantum consciousness, though these remain speculative and highly debated within the scientific community.

$$H = \sum_i E_i |i\rangle\langle i| + \sum_{i \neq j} J_{ij} \left(|i\rangle\langle j| + |j\rangle\langle i| \right) \tag{3.30}$$

Equation 3.30 represents a Hamiltonian describing interactions between neurons, where E_i are energy levels of neural states $|i\rangle$ and J_{ij} are coupling constants between different states. This model could be extended to quantum neural networks, allowing for the exploration of how quantum effects might influence neural processing.

Training DNNs is a computationally intensive process that involves optimizing complex functions across high-dimensional spaces. Quantum computing, with its ability to perform parallel computations and explore multiple solutions simultaneously, holds the promise of dramatically improving this process. Consider the task of training a neural network for NLP, a field that requires vast amounts of data and computational resources. Quantum computers could accelerate this training process, leading to more powerful AI systems capable of understanding and generating human language with greater nuance and accuracy.

Quantum annealing, for example, exploits quantum tunneling to escape local minima in optimization landscapes, potentially finding global minima more efficiently than classical methods. This capability could be crucial in training neural networks that underpin AGI, enabling them to learn more effectively and solve problems that are currently beyond the reach of classical AI.

$$H = -\Gamma(t)\sum_i \sigma_i^x + H_{\text{problem}}(\sigma_i^z) \tag{3.31}$$

In Equation 3.31, the Hamiltonian H represents the quantum system used in annealing, where $\Gamma(t)$ controls quantum tunneling, and H_{problem} encodes the optimization problem. This process enables the exploration of potential solutions in a high-dimensional space, facilitating the training of complex neural networks. An anecdotal example could be D-Wave's quantum annealing system, which has been applied to solve optimization problems in logistics and finance, illustrating the practical benefits of this technology even in its early stages.

Quantum computing has the potential to revolutionize the pursuit of AGI by providing the computational power needed to emulate the brain's complex processes. By leveraging quantum mechanics' principles—such as superposition, entanglement, and quantum interference—quantum computing can enhance learning, optimize neural networks, and simulate cognitive functions with unprecedented efficiency.

Integrating quantum computing into AI will likely play an increasingly central role in developing intelligent systems capable of achieving AGI. The synergy between quantum physics and AI promises to push the boundaries of what is computationally possible, opening new frontiers in our understanding of intelligence and the brain. Imagine a future where quantum computers not only power the next generation of AI but also help us unravel the mysteries of consciousness itself, leading to breakthroughs that could redefine what it means to be human.

Innovative Architectures

The journey toward AGI has been marked by significant progress in machine learning, deep learning, and neural networks. However, in isolation, these technologies have inherent limitations that prevent them from fully realizing the vision of AGI. For instance, deep learning models, while powerful in pattern recognition, cannot often perform abstract reasoning or transfer knowledge efficiently across different domains. Similarly, traditional symbolic AI can handle structured problem-solving and logical reasoning but struggles with learning from unstructured data or with adapting to new, unforeseen scenarios [36].

These limitations highlight the need for innovative architectures in the pursuit of AGI. Such architectures aim to transcend the boundaries of individual AI approaches by synergistically combining their strengths, thereby paving the way for more versatile and adaptable AI systems. By addressing the weaknesses of standalone models and leveraging their complementary capabilities, innovative architectures offer a promising path toward achieving the holistic, integrated form of intelligence that AGI represents.

Hybrid Models: Combining Neural and Symbolic AI

The evolution of AI has seen the development of two primary paradigms: neural-based approaches and symbolic AI. These paradigms, each with its unique strengths and limitations, have shaped the trajectory of AI research and application. Neural-based approaches, particularly deep learning, excel in recognizing patterns, making inferences from large datasets, and adapting to new information. These methods have achieved remarkable success in tasks such as image recognition, natural language processing, and autonomous driving, where the ability to learn from vast amounts of data is crucial.

Symbolic AI, on the other hand, operates on a foundation of logical rules and structured knowledge representation. This approach emphasizes the explicit encoding of human knowledge in the form of symbols, rules, and relationships, allowing for robust reasoning, explainability, and manipulation of abstract concepts. Symbolic AI has been instrumental in areas such as expert systems, where the ability to model and reason about complex domains with precision and clarity is paramount.

However, both approaches have inherent limitations. Despite their prowess in pattern recognition, neural networks often struggle with interpretability, reasoning, and the ability to incorporate structured knowledge. This has led to the perception of neural networks as "black boxes," where the decision-making process is opaque and difficult to understand. Additionally, neural networks can be data-hungry, requiring vast amounts of labeled data for training, which may not always be available.

While offering transparency and reasoning capabilities, symbolic AI lacks the adaptability and learning flexibility that neural networks provide. Symbolic systems are typically rigid, requiring manual updates to their rule sets and struggling to cope with the variability and complexity inherent in real-world data. They are often brittle, failing when confronted with scenarios not explicitly accounted for in their rule-based framework.

Recognizing these two approaches' complementary strengths and weaknesses, researchers have explored hybrid models that combine neural and symbolic AI to harness the best of both worlds [37]. These hybrid models aim to integrate the intuitive pattern recognition capabilities of neural networks with the explicit reasoning, explainability, and structured knowledge manipulation of symbolic AI.

One prominent example of hybrid AI is the neuro-symbolic model, which integrates neural networks with symbolic reasoning systems. In this model, neural networks are used to process raw sensory input, such as images or text, extracting relevant features and patterns. These features are then passed to a symbolic reasoning module, which uses logical rules and structured knowledge to make decisions, draw inferences, and generate explanations.

$$\text{Hybrid Objective: } \mathcal{L}_{\text{hybrid}} = \mathcal{L}_{\text{neural}} + \lambda \mathcal{L}_{\text{symbolic}} \tag{3.32}$$

In Equation 3.32, L_{neural} represents the loss associated with the neural network component, L_{symbolic} represents the loss associated with the symbolic reasoning component, and λ is a weighting factor that balances the contributions of both components to the overall model.

One application of this hybrid approach can be seen in neuro-symbolic visual question answering (VQA). In this task, a neural network processes an image to extract visual features, such as the presence of objects, their relationships, and spatial configurations. These features are then used by a symbolic reasoning system to answer complex questions about the image, such as "What is the object to the left of the red ball?" The symbolic component enables the model to reason about spatial relationships and object identities explicitly, leading to more accurate and interpretable answers.

Another example of hybrid AI is found in knowledge graphs, where symbolic knowledge representations are combined with neural embeddings to capture both the structured and unstructured aspects of information. Knowledge graphs represent entities and their relationships in a structured form, making them ideal for tasks that require reasoning over complex relationships, such as recommendation systems or semantic search. Neural networks are used to embed these entities and relationships into continuous vector spaces, allowing for efficient computation and generalization to new, unseen data.

$$\text{Knowledge Graph Embedding: } \mathbf{h}_{\text{entity}} = f\left(\text{symbolic knowledge, neural embedding}\right) \tag{3.33}$$

In Equation 3.33, $\mathbf{h}_{\text{entity}}$ represents the embedding of an entity in the knowledge graph, which is computed using both symbolic knowledge (such as logical relationships) and neural embeddings (which capture patterns and similarities from data).

The combination of neural and symbolic approaches in hybrid models offers several advantages:

- **Enhanced Interpretability:** The symbolic component of hybrid models provides explicit reasoning and explanations, making the decision-making process more transparent and understandable.
- **Improved Generalization:** Neural networks contribute their powerful pattern recognition capabilities, enabling the model to generalize from data and handle variability and noise.

- **Structured Knowledge Integration:** Symbolic AI allows the model to incorporate structured knowledge, such as ontologies or rules, enabling more robust reasoning and decision-making.
- **Flexibility and Adaptability:** The neural component allows the model to adapt to new data and learn from experience, while the symbolic component ensures that the model adheres to logical consistency and domain-specific rules.

Despite their potential, hybrid models also face challenges. Integrating neural and symbolic components requires careful balancing because the two approaches operate on different principles and representations. Moreover, developing efficient training algorithms that can optimize both components simultaneously remains an area of active research. Nevertheless, the promise of hybrid AI lies in its ability to combine the strengths of both neural and symbolic approaches, paving the way for more versatile and capable AI systems.

Modular Architectures: Building Flexible AI Systems

In the pursuit of AGI, the design of AI systems that are adaptable, scalable, and capable of handling a wide array of tasks with humanlike flexibility is paramount. Modular architectures offer a promising solution by proposing that AI systems be composed of distinct, specialized components—or modules—that can be integrated and reconfigured as needed. This approach mirrors the modular nature of the human brain, where different regions are specialized for various cognitive functions yet work together seamlessly to produce coherent behavior. The concept of modularity in AI is rooted in the idea that complex tasks can often be decomposed into simpler sub-tasks, each of which can be managed by a specialized module. For instance, in natural language processing, separate modules might be designed for tasks such as syntactic parsing, semantic understanding, and text generation. By dividing the problem space, each module can be optimized for its specific function, leading to greater overall efficiency and effectiveness.

Mathematically, modular architectures can be understood as systems of interconnected functions, where each function f_i represents a module processing a specific aspect of the input data x. The output of the system is then a composition of these functions, expressed as:

$$y = f_n\left(f_{n-1}\left(\ldots f_2\left(f_1\left(x\right)\right)\ldots\right)\right) \tag{3.34}$$

This layered composition not only enhances the system's ability to manage complexity but also allows for modular updates—where individual functions can be replaced or improved without altering the entire system. One of the primary advantages of modular architectures is their flexibility. In a modular system, new modules can be added to handle new tasks without requiring a complete redesign of the system. This is particularly critical for AGI, where the range of tasks the system might need to perform is vast and varied. Just as the brain can learn new skills and adapt to new environments by recruiting and reconfiguring different neural circuits, a modular AI system can be extended and adapted by integrating new modules or reconfiguring existing ones.

Moreover, modularity supports scalability. As the demands on an AI system grow, additional modules can be incorporated to meet these demands. This approach allows the system to scale in both functionality and complexity without becoming unwieldy or difficult to manage. In contrast, monolithic AI systems, which attempt to handle all tasks within a single, integrated framework, often struggle with scalability, because increasing complexity can lead to inefficiencies and reduced performance.

The modular nature of the brain provides a compelling biological analogy. The brain's modular structure allows for parallel processing and specialization, where different regions handle tasks such as vision, language, and motor control. This specialization enables the brain to operate efficiently and adapt to new challenges, much as how a modular AI system can be designed to handle a variety of tasks through specialized components.

Another critical aspect of modular architectures is their potential for fostering innovation. By allowing different teams or researchers to focus on developing and improving individual modules, modular systems encourage experimentation and innovation within specific areas of AI. These innovations can then be integrated into the larger system, enhancing its overall capabilities. This approach is akin to the way different areas of neuroscience focus on specific aspects of brain function, with findings in one area often informing and enhancing our understanding of others.

In addition, modular architectures align well with the current trend toward interdisciplinary research in AI and cognitive science. The development of AGI is not just a challenge for computer scientists; it requires insights from psychology, neuroscience, linguistics, physics, biology, chemistry, and other fields. By structuring AI systems in a modular way, researchers from different disciplines can contribute their expertise to specific modules, which can then be integrated into a coherent whole. This interdisciplinary approach mirrors the way different scientific fields contribute to our understanding of the brain, with each discipline shedding light on different aspects of cognition.

While the benefits of modular architectures are clear, there are also challenges associated with this approach. One of the main challenges is ensuring seamless integration between modules. In the human brain, different regions are highly interconnected, with communication between them occurring at both a local and a global level. Replicating this level of integration in AI systems is a complex task. Modules must be designed to work together harmoniously, with clear protocols for communication and data exchange. Without careful design, a modular system can become fragmented, with modules failing to cooperate effectively or even working at cross-purposes.

Another challenge is the potential for redundancy and inefficiency. In a modular system, there is a risk that different modules might duplicate functions or fail to share information efficiently. This can lead to wasted resources and reduced overall performance. Addressing this challenge requires careful coordination and standardization, ensuring that modules complement rather than compete with one another. The brain addresses this issue through mechanisms like feedback loops and hierarchical organization, where higher-level processes coordinate and integrate the functions of lower-level ones. Developing analogous mechanisms in AI systems is an active area of research.

To enhance modular architectures further, the inclusion of a central coordination module—akin to the brain's prefrontal cortex—can be considered. This central module would be responsible for managing the interactions between specialized modules, ensuring that their outputs are synthesized into a coherent whole. Such a structure would be essential for tasks requiring integration across different domains, much like how the human brain integrates sensory inputs, memory, and reasoning to make decisions.

Despite the challenges, the modular approach holds great promise for advancing AGI. By drawing inspiration from the brain's modular structure, AI researchers can design systems that are not only more flexible and scalable but also more robust and capable of adapting to new challenges. As research in modular architectures continues, we may discover that they offer the key to unlocking the full potential of AGI, leading to systems that are as versatile, resilient, and intelligent as the human brain itself [38].

Quantum-Hybrid Architectures: Beyond Classical Computing

As AI continues to push the boundaries of what is computationally possible, the limitations of classical computing architectures are becoming increasingly apparent. Classical systems, despite their remarkable achievements, are fundamentally constrained by their reliance on binary logic and traditional von Neumann architectures. These constraints are particularly evident when tackling problems that involve vast combinatorial spaces, complex optimization tasks, or the simulation of quantum systems themselves—areas where classical approaches may falter due to their exponential scaling in computational complexity.

Quantum computing, with its foundations in the principles of quantum mechanics, offers a revolutionary approach to these challenges. Quantum computers operate on qubits, which, unlike classical bits, can exist in superpositions of states, allowing them to process a vast amount of information simultaneously. This inherent parallelism enables quantum computers to explore multiple solutions at once, making them exceptionally powerful for specific types of problems, particularly those involving optimization, cryptography, and the simulation of quantum phenomena.

In the context of AI, quantum-hybrid architectures emerge as a promising avenue for integrating quantum computing with classical AI frameworks. These architectures combine the strengths of quantum and classical computing to create systems capable of tackling problems that are currently intractable for classical systems alone. The key idea is to use quantum computing to perform specific sub-tasks within an AI system—tasks that benefit from quantum speed-up—while leveraging classical computing for other aspects that are better suited to traditional architectures.

One of the primary motivations for quantum-hybrid architectures is the potential to revolutionize machine learning, particularly in the training and inference stages of deep learning models. Quantum machine learning (QML) leverages quantum algorithms to enhance these processes. For instance, quantum versions of classical algorithms, such as the QAOA or the variational quantum eigensolver (VQE), have shown promise in solving optimization problems more efficiently than their classical counterparts. These algorithms are particularly useful in the training of models, where finding optimal weights and parameters often involves navigating a highly complex landscape.

The potential of quantum-hybrid architectures can be mathematically framed by considering the quantum-enhanced version of a typical optimization problem in AI. Suppose a classical neural network aims to minimize a loss function $L(\theta)$ with respect to its parameters θ. In a quantum-hybrid architecture, a quantum co-processor could be employed to explore the parameter space more efficiently. The quantum-assisted optimization process can be represented as:

$$\theta_{\text{opt}} = \text{argmin}_{\theta} \langle \psi(\theta) | \hat{H} | \psi(\theta) \rangle + L(\theta) \tag{3.35}$$

Here, $\langle \psi(\theta) | \hat{H} | \psi(\theta) \rangle$ represents the quantum expectation value of a Hamiltonian $\hat{H}$, which encodes the problem's constraints, and $L(\theta)$ represents the classical loss function. The hybrid approach allows the system to leverage quantum mechanics to explore potential solutions, while classical methods refine and finalize the optimization.

Beyond optimization, quantum computing also offers new paradigms for data representation and processing. Quantum data, represented by qubits, can encode and process information in ways that classical systems cannot. For example, quantum entanglement and superposition allow for the creation of complex, nonclassical correlations that could be used to capture intricate patterns in data, offering a new perspective on feature extraction and data manipulation.

Quantum-hybrid architectures also present exciting possibilities for advancing the simulation of quantum systems, a task that is notoriously difficult for classical computers. In material science, chemistry, and physics, simulating the behavior of molecules, materials, and other quantum systems requires enormous computational resources due to the exponential scaling of the quantum state space with the number of particles. Quantum computers are inherently suited to this task because they operate according to the same principles that govern the systems being simulated. By integrating quantum simulators into AI systems, researchers can develop models that not only simulate quantum systems more efficiently but also use these simulations to inform AI-driven predictions and decisions in real time.

Another promising application of quantum-hybrid architectures is in the realm of cryptography and security. Quantum computers are known for their potential to break classical cryptographic systems, such as RSA (Rivest-Shamir-Adleman), through algorithms like Shor's algorithm. However, they also offer the possibility of creating new, quantum-resistant cryptographic protocols. In AI, where data security and privacy are paramount, quantum-hybrid architectures could be employed to develop and implement these new protocols, ensuring that AI systems remain secure against both classical and quantum threats.

Despite the enormous potential, the integration of quantum computing into AI systems is not without challenges. Quantum computers are still in their early stages of development, with current quantum devices—often referred to as noisy intermediate-scale quantum (NISQ) computers—being limited by noise, decoherence, and error rates. These limitations restrict the size and complexity of problems that can currently be tackled. However, ongoing advancements in quantum error correction, qubit fidelity, and quantum algorithms are gradually overcoming these hurdles, bringing the vision of practical quantum-hybrid architectures closer to reality.

Furthermore, the successful implementation of quantum-hybrid architectures requires the development of sophisticated hybrid algorithms and software frameworks that can seamlessly integrate quantum and classical computing. This integration involves creating interfaces that allow classical systems to offload specific tasks to quantum processors and then reintegrate the results into the broader AI system. The development of such frameworks is a vibrant area of research, with initiatives such as quantum-classical programming languages, quantum software development kits, and cloud-based quantum computing platforms paving the way for broader adoption.

The implications of quantum-hybrid architectures extend beyond AI to the broader field of computing. By introducing quantum computing into the AI workflow, researchers can explore new computational paradigms that challenge conventional approaches. These architectures not only offer a pathway to overcoming current computational limits but also provide a fertile ground for discovering novel algorithms, data structures, and methods that could redefine the future of computing itself [39].

Evolving Architectures: The Role of Evolutionary Algorithms

In the quest for AGI, evolving architectures represent a dynamic and adaptive methodology, drawing inspiration from the principles of natural evolution to guide the development of AI systems. Unlike traditional AI approaches that rely on manual design and optimization, evolving architectures utilize evolutionary algorithms to automatically generate, select, and refine models and systems. This approach mirrors the process of natural selection, where the fittest individuals survive and reproduce, passing their advantageous traits to the next generation. By applying this concept to AI, evolutionary algorithms offer a powerful tool for exploring a vast space of potential architectures, identifying those that are most effective for given tasks.

Evolutionary algorithms are rooted in the field of evolutionary computation, a branch of artificial intelligence that applies principles such as selection, mutation, and crossover—key mechanisms of natural evolution—to optimize solutions to complex problems. These algorithms typically operate on a population of candidate solutions, each represented by a set of parameters (often called a genome). Over successive generations, the population evolves as the algorithms select the most promising candidates based on a fitness function, which evaluates their performance on specific tasks.

The process of evolving AI architectures involves several key steps, which can be mathematically formalized as follows:

1. **Initialization:** A population of candidate architectures is generated, often randomly. Each candidate A_i in the population P_0 is represented by a set of parameters θ_i.
2. **Evaluation:** The fitness of each candidate architecture is evaluated using a fitness function $F(A_i)$, which measures how well the architecture performs on a given task. This evaluation could involve training a neural network and assessing its accuracy, generalization, or efficiency.
3. **Selection:** The most fit candidates are selected to form a new population. This selection process can be based on various strategies, such

as tournament selection or rank-based selection, where the probability of selection is proportional to fitness.

4. **Crossover and Mutation:** The selected candidates undergo crossover and mutation operations to create offspring. Crossover combines the parameters of two parent architectures to produce new candidates, while mutation introduces random changes to parameters, promoting diversity in the population.
5. **Iteration:** The process is repeated for multiple generations, with each iteration producing increasingly refined and optimized architectures.

Mathematically, the evolution of architectures can be expressed as:

$$P_{t+1} = \text{Selection}\left(\text{Crossover}\left(\text{Mutation}\left(P_t\right)\right)\right) \tag{3.36}$$

where P_t is the population at generation t, and the operations of selection, crossover, and mutation are applied to generate the population P_{t+1} for the next generation.

Evolving architectures offer several advantages in the context of AGI. One of the most significant benefits is their ability to discover novel architectures that human designers might not conceive. By exploring a vast space of possible configurations, evolutionary algorithms can identify architectures that are not only effective but also efficient and robust, often outperforming hand-designed models. This exploratory capability is particularly valuable in the pursuit of AGI, where the optimal architecture may be unknown or too complex to design manually.

Another advantage of evolving architectures is their adaptability. As AI systems are deployed in dynamic environments, they must be capable of adapting to new challenges and data. Evolutionary algorithms enable this adaptability by continuously refining architectures based on feedback from the environment, much like how biological organisms evolve in response to changing conditions. This adaptability is crucial for developing AI systems that can operate effectively across a wide range of tasks and scenarios, a key requirement for AGI.

Furthermore, evolving architectures offer a pathway to reducing the reliance on large-scale data and computational resources. Traditional AI approaches, particularly those based on deep learning, often require massive datasets and extensive computational power to train effective models. In contrast, evolutionary algorithms can optimize architectures with fewer resources by focusing on the quality of solutions rather than the quantity of data. This efficiency can be particularly advantageous in situations where data is scarce or expensive to obtain.

However, the application of evolutionary algorithms to AI development also presents challenges. One of the primary difficulties is the computational cost of evaluating candidate architectures, especially in complex domains where training and testing a model can be time-consuming. To address this, researchers have developed various strategies to reduce the computational burden, such as using surrogate models to approximate the fitness function or employing parallel and distributed computing to evaluate multiple candidates simultaneously.

Another challenge is ensuring the scalability of evolving architectures. As the complexity of AI systems increases, the search space of possible architectures grows exponentially, making it difficult to explore effectively. To mitigate this, researchers often combine evolutionary algorithms with other optimization techniques, such as gradient-based methods or reinforcement learning, to guide the search process more efficiently.

In the broader context of AI research, evolving architectures also contribute to our understanding of the relationship between evolution and intelligence. By simulating evolutionary processes in a computational setting, these architectures offer insights into how complex cognitive systems might emerge and evolve over time. This understanding could inform the design of more advanced AI systems and shed light on the evolutionary processes that have shaped human intelligence [40].

Integrating Diverse Approaches for AGI

The pursuit of AGI necessitates a convergence of multiple disciplines, each contributing unique perspectives and methodologies toward the creation of a truly intelligent system. While individual AI techniques, such as deep learning, symbolic reasoning, and evolutionary algorithms, have made significant strides, the integration of these diverse approaches holds the key to developing a system that mirrors the complexity and adaptability of human cognition.

One of the primary challenges in achieving AGI is the fragmented nature of current AI research. Many successful AI applications are highly specialized, excelling in specific tasks like image recognition, natural language processing, or game playing, but these systems often lack the generalization capability necessary for broader, more humanlike intelligence. To address this, researchers are increasingly focused on creating hybrid models that combine the strengths of different AI paradigms, enabling systems to learn, reason, and adapt across a wide range of domains.

Integrating neural networks with symbolic reasoning is one such approach that has gained traction. Neural networks, particularly deep learning models, are adept at handling unstructured data and recognizing patterns, but they often struggle with tasks requiring explicit reasoning, logic, and understanding of abstract concepts. On the other hand, symbolic AI, which dominated the early days of AI research, excels in rule-based reasoning and knowledge representation but lacks the flexibility and scalability of neural networks. By combining these paradigms, researchers aim to create systems that can not only learn from vast amounts of data but also apply logical reasoning and problem-solving skills in novel situations.

A practical implementation of this integration can be seen in neurosymbolic systems, where neural networks are used for perception and data interpretation, while symbolic reasoning modules handle tasks that require structured knowledge and logic. This dual-system approach allows for more robust decision-making, where the strengths of each component complement the other, leading to a more versatile and capable AI.

Another area of integration involves the blending of data-driven and model-based approaches. While data-driven methods, such as machine learning, rely heavily on large datasets to extract patterns and make predictions, model-based approaches use mathematical models to simulate and predict the behavior of systems based on underlying principles. Combining these approaches can enhance the ability of AI systems to operate in environments with limited data or where understanding the underlying mechanisms is crucial. For example, in scientific discovery, AI systems that integrate machine learning with physical models can accelerate research by generating hypotheses, conducting simulations, and refining models based on experimental data.

The integration of evolutionary algorithms with other AI methodologies also offers significant promise. Evolutionary algorithms, inspired by natural selection, are powerful tools for exploring a vast search space of potential solutions, optimizing complex systems, and evolving architectures over time. When combined with deep learning or reinforcement learning, evolutionary algorithms can help discover novel network architectures, optimize hyperparameters, or evolve strategies for decision-making in dynamic environments.

Furthermore, advances in cognitive science and neuroscience are increasingly informing AI research, particularly in the development of cognitive architectures that seek to emulate the human mind's structure and function. By integrating insights from these fields, AI systems can be designed to mirror the processes of human learning, memory, and problem-solving, leading to more natural and intuitive interactions between humans and machines. For instance, cognitive architectures like ACT-R and SOAR incorporate elements of human cognition, such as attention, working memory, and procedural knowledge, offering a more holistic approach to AI design.

Quantum computing also presents a frontier for the integration of diverse approaches in AI. With their ability to perform complex computations in parallel, quantum computers have the potential to revolutionize AI by enabling the processing of vast amounts of data and the simulation of complex systems at unprecedented speeds. Integrating quantum computing with existing AI methodologies could open new possibilities for solving problems that are currently intractable for classical computers, such as large-scale optimization, cryptography, and the simulation of quantum systems themselves.

The multidisciplinary nature of AGI research underscores the importance of collaboration across fields. Neuroscientists, cognitive scientists, computer scientists, mathematicians, and engineers must work together to create systems that not only replicate human intelligence but also extend its capabilities. This collaborative effort requires a shared understanding of the principles underlying human cognition, the development of new computational paradigms, and the exploration of how these diverse approaches can be synthesized into a unified framework for AGI.

In integrating these approaches, researchers must also consider the ethical implications and societal impact of AGI. As AI systems become more powerful and autonomous, ensuring their alignment with human values and goals becomes increasingly critical. The integration of ethical reasoning, transparency, and explainability into AI systems is essential for building trust and ensuring that AGI serves the broader good of humanity.

This synthesis of diverse approaches is not just a technical challenge but a philosophical one, as it forces researchers to confront fundamental questions about the nature of intelligence, consciousness, and what it means to create a machine that can think, learn, and adapt like a human. The integration of these varied perspectives will likely be the defining challenge of AGI research in the coming decades, requiring not just advances in technology, but a deeper understanding of the human mind and the nature of intelligence itself.

Cross-Disciplinary Integration

Cross-disciplinary integration in AI research represents a vital strategy for advancing toward AGI, reflecting the multifaceted nature of human cognition. Human intelligence is a product of the intricate interplay of various cognitive functions—ranging from perception and memory to reasoning and emotional processing. Similarly, the realization of AGI will likely hinge on the successful synthesis of diverse AI methodologies, including neural-inspired learning, symbolic reasoning, sensory perception, emotional intelligence, and more. This integration is not merely a technical necessity but also a conceptual one, as it allows for a richer understanding of intelligence and the development of systems that are capable of humanlike adaptability and problem-solving.

The value of cross-disciplinary integration lies in its ability to facilitate the borrowing of insights and methodologies from one domain to address challenges in another. For example, insights from cognitive science can inform the development of more effective learning algorithms, while advances in neuroscience can inspire the creation of architectures that better mimic the structure and function of the human brain. This exchange of ideas across disciplines spurs innovation and accelerates the path to AGI by providing a more holistic understanding of intelligence and its underlying mechanisms.

In practice, cross-disciplinary integration can be seen in efforts to combine neural networks with symbolic reasoning systems. Neural networks excel at pattern recognition and learning from unstructured data, but they often lack the explicit reasoning and interpretability that symbolic systems offer. By integrating these two approaches, researchers aim to develop AI systems that can not only learn from data but also reason about the world in a manner akin to human thought. This fusion allows for more robust AI systems that are capable of handling a wider range of tasks with greater accuracy and reliability.

Another promising area of integration is the combination of sensory perception with emotional intelligence. While AI systems have made great strides in processing sensory data—such as images, sound, and text—integrating these capabilities with an understanding of emotional and social cues remains a significant challenge. Emotional intelligence, which involves recognizing, understanding, and responding to human emotions, is a critical component of human cognition and interaction. Incorporating this dimension into AI systems could lead to more natural and effective human-machine interactions, particularly in areas such as healthcare, customer service, and education.

Despite its potential, the integration of different AI technologies presents significant challenges. One of the foremost challenges is compatibility. Different AI approaches often rely on distinct representations of knowledge, processing mechanisms, and learning paradigms, making seamless integration a complex task. For example, neural networks and symbolic systems represent knowledge in fundamentally different ways—neural networks through distributed representations and symbolic systems through discrete, rule-based structures. Bridging these differences requires innovative solutions that can harmonize these disparate approaches while preserving their unique strengths.

Scalability is another major challenge in cross-disciplinary integration. As AI systems become more complex, combining multiple technologies can exponentially increase the computational resources required. This is particularly true when integrating deep learning models, which are already computationally intensive, with other AI approaches. Addressing scalability requires not only advances in hardware and computational efficiency but also the development of more efficient algorithms that can manage the increased complexity without sacrificing performance.

Furthermore, maintaining a balance between the interpretability provided by symbolic systems and the powerful data-driven insights offered by neural approaches is crucial. One of the key advantages of symbolic AI is its transparency and explainability—qualities that are often lacking in deep learning models, which are frequently criticized as "black boxes." Ensuring that integrated AI systems are both powerful and understandable is essential for their adoption in critical applications where trust and accountability are paramount, such as in healthcare, finance, and autonomous systems.

The challenge of cross-disciplinary integration extends beyond technical considerations to include ethical and societal implications. As AI systems become more capable and autonomous, ensuring that they align with human values and can be trusted to make decisions in complex, real-world environments is increasingly important. This requires the integration of ethical reasoning into AI systems, allowing them to consider the broader impact of their actions and decisions. Additionally, cross-disciplinary integration must account for the diverse needs and perspectives of different stakeholders, including researchers, policymakers, and the public, to ensure that AGI development proceeds in a manner that is inclusive and beneficial for all.

Unified Model of AGI

The quest for AGI envisions a unified model that encapsulates the full spectrum of human cognitive abilities within a single, coherent framework. Such a model would not only replicate the individual components of human intelligence—such as perception, memory, reasoning, and language processing—but also integrate these capabilities into a system that can operate with the same fluidity and adaptability as the human mind. This ambitious goal requires a multidisciplinary approach that combines insights from artificial intelligence, cognitive science, neuroscience, and other related fields.

To develop a unified model of AGI, one must consider the diverse ways in which humans process information and interact with the world. For instance, human learning is not limited to any single method but involves a combination of supervised learning (learning from labeled examples), unsupervised learning (identifying patterns in unlabeled data), reinforcement learning (learning from trial and error), and transfer learning (applying knowledge from one domain to another). A unified AGI model would need to incorporate all these learning paradigms, enabling it to learn from experience, adapt to new environments, and generalize knowledge across different tasks.

In addition to learning, the unified model must also encompass human-like reasoning and problem-solving abilities. This includes both deductive reasoning (drawing specific conclusions from general principles) and inductive reasoning (deriving general principles from specific observations). The model must be capable of handling abstract concepts and applying them to concrete situations, much as humans can reason about hypothetical scenarios and make decisions based on incomplete or ambiguous information. Furthermore, it should be able to engage in commonsense reasoning, understanding the everyday physical and social world in a way that aligns with human intuition. Natural language understanding and generation are also critical components of the unified AGI model. Language is not only a tool for communication but also a medium for thought, enabling humans to express complex ideas, share knowledge, and collaborate on solving problems. The model must be able to comprehend and generate natural language across various contexts, ranging from casual conversation to technical discourse. This requires a deep understanding of semantics, syntax, pragmatics, and the nuances of human communication, as well as the ability to generate coherent and contextually appropriate responses.

Emotion recognition and expression are also essential features of the unified AGI model. Emotions play a vital role in human cognition, influencing decision-making, social interactions, and motivation. The model must be capable of recognizing and interpreting emotional cues from both verbal and nonverbal signals, such as facial expressions, tone of voice, and body language. Additionally, it should be able to express emotions in a manner that is appropriate to the context, fostering more natural and empathetic interactions with humans.

The ability to interact with the physical world is another crucial aspect of the unified model. This includes not only the manipulation of objects through robotic systems but also the perception of the environment through sensors and the interpretation of sensory data. The model must be able to navigate and interact with the physical world in real time adapting its behavior based on the dynamics of its surroundings. This requires the integration of advanced robotics, computer vision, and sensor fusion technologies, allowing the AGI system to operate autonomously in diverse and unpredictable environments.

However, the development of a unified AGI model is fraught with challenges, one of the most pressing being the *alignment problem*. The alignment problem concerns ensuring that the goals and behaviors of AGI systems are consistent with human values and intentions. As AGI systems gain greater autonomy and decision-making capabilities, there is an inherent risk that their objectives may diverge from those of their creators, leading to outcomes that could be detrimental to humanity. This misalignment could manifest in subtle ways, such as optimizing for objectives that are technically correct but ethically or socially undesirable, or in more extreme scenarios, where an AGI system could pursue goals that pose existential risks.

One approach to addressing the alignment problem is the concept of coherent extrapolated volition (CEV), which suggests that AGI should be designed to act in accordance with the collective will of humanity, as it would be if humans were more informed, rational, and in better agreement. The idea is that rather than programming AGI with a fixed set of values, we would design it to extrapolate and respect the values and preferences that humans would want if they had the opportunity to think things through more thoroughly. Implementing CEV or similar frameworks could help ensure that AGI systems make decisions that align with human values and contribute to the well-being of society.

Philosophical considerations are also integral to the pursuit of a unified AGI model. Understanding the nature of intelligence, consciousness, and self-awareness is essential for creating systems that not only mimic human cognition but also possess a form of agency and autonomy. These questions touch on the very essence of what it means to be intelligent and conscious, and addressing them will require a collaborative effort across disciplines, bringing together philosophers, cognitive scientists, and AI researchers.

The development of a unified model of AGI will likely be an iterative process, characterized by continuous advancements in technology and our understanding of intelligence. Each iteration will build upon the previous one, incorporating new discoveries and refining existing approaches. The integration of emerging technologies, such as quantum computing and brain-inspired architectures, may provide the computational power and design principles necessary to achieve this vision.

As the development of a unified AGI model progresses, it is imperative to consider the ethical implications and societal impact of such systems. Ensuring that AGI aligns with human values and operates within ethical boundaries is not only a technical challenge but also a moral responsibility. This requires the development of frameworks and guidelines that govern the behavior of AGI systems, ensuring that they act in ways that are beneficial to humanity as a whole. By addressing these challenges head-on, we can strive toward creating AGI systems that enhance our world and contribute to the flourishing of human civilization.

Pioneering the Path to Artificial General Intelligence

As we conclude this chapter on the pursuit of AGI, we find ourselves at the intersection of human ingenuity and technological evolution. The path toward AGI is not merely a technical endeavor but a profound journey into the very nature of intelligence, consciousness, and what it means to be human. Over the past several decades, advancements in machine learning, deep learning, and natural language processing have brought us closer to this elusive goal, yet the true realization of AGI remains one of the most formidable challenges of our time.

The ambition behind AGI is to create systems that possess a level of cognitive flexibility and adaptability that rivals human intelligence. Unlike narrow AI systems, which excel at specific tasks such as image recognition or language translation, AGI aspires to understand, learn, and reason across a broad array of domains, much like the human mind. This chapter has explored the theoretical underpinnings, technological innovations, and philosophical implications of AGI, presenting a roadmap that spans the convergence of various AI methodologies and the integration of humanlike cognitive processes within machines.

The journey toward AGI is as much about understanding ourselves as it is about building intelligent machines. Human cognition is a complex interplay of perception, memory, learning, reasoning, and emotion, all of which contribute to our ability to navigate the world with nuance and adaptability. In aspiring to replicate these abilities within a machine, we are, in effect, creating a mirror that reflects our deepest questions about the mind, consciousness, and the essence of intelligence.

Throughout this chapter, we have discussed the necessity of a unified model of AGI—one that seamlessly integrates diverse cognitive functions into a coherent whole. This model must not only mimic the individual components of human intelligence but also replicate the dynamic interactions between them. The fluidity with which humans transition from perceiving a situation to reasoning about it, drawing on memories, learning from the experience, and applying this knowledge to new contexts is a hallmark of our cognitive architecture. Capturing this in a machine requires a multidisciplinary approach that draws on insights from artificial intelligence, cognitive science, neuroscience, and even philosophy.

One of the most significant challenges in developing AGI is achieving a balance between learning efficiency and generalization. Human beings have an extraordinary ability to learn from a small number of examples and to apply knowledge across different domains—something current AI systems struggle with. Developing algorithms that can learn efficiently from limited data and generalize this knowledge to new, unseen situations is a critical step toward AGI. This chapter has outlined various learning paradigms—supervised, unsupervised, reinforcement, and transfer learning—that must be integrated into AGI systems to achieve this level of adaptability.

Another critical aspect of AGI is the ability to reason and solve problems in a humanlike manner. Human reasoning is not a linear process but a complex interplay of deductive, inductive, and abductive reasoning, often influenced by emotions, biases, and social contexts. An AGI system must be capable of handling abstract concepts, reasoning about hypothetical scenarios, and making decisions based on incomplete or ambiguous information. This requires not only advanced computational models but also a deep understanding of human cognition and how it can be translated into machine algorithms.

Language plays a pivotal role in human thought and communication, making natural language understanding and generation a cornerstone of AGI. The ability to comprehend and generate language across various contexts—whether casual conversation, technical discourse, or creative writing—is essential for AGI to interact with humans meaningfully. The challenge lies not only in processing the syntax and semantics of language but also in understanding the pragmatics, context, and subtleties that make human communication so rich and nuanced. This chapter has explored the advancements in natural language processing and how they contribute to the broader goal of AGI.

Emotions, often overlooked in discussions about intelligence, are integral to human cognition. They influence our decision-making, shape our social interactions, and motivate our actions. For AGI to be truly humanlike, it must recognize and respond to emotions, both in itself and in others. This involves developing systems that can interpret emotional cues from verbal and nonverbal signals and express emotions in a way that is contextually appropriate and empathetic. The integration of emotion recognition and expression into AGI is not just a technical challenge but also a philosophical one, raising questions about the nature of artificial consciousness and whether machines can ever truly "feel." The physical embodiment of AGI in the world is another significant challenge. Humans interact with their environment through a complex sensory-motor system that allows them to perceive, manipulate, and navigate the physical world. Replicating this in an AGI system requires advances in robotics, computer vision, and sensor fusion, enabling machines to operate autonomously in real-world environments. This chapter has highlighted the importance of developing AGI systems that can perceive their surroundings, interpret sensory data, and make decisions in real time, adapting to the dynamics of their environment.

While the technical challenges of AGI are immense, they are matched by equally significant ethical and societal concerns. The potential impact of AGI on society cannot be overstated. A machine with human-level intelligence could revolutionize industries, transform economies, and even reshape our daily lives. However, with such transformative power comes the responsibility to ensure that AGI is developed and deployed in ways that align with human values and promote the common good.

The alignment problem—ensuring that AGI systems act in accordance with human intentions and values—is one of the most pressing challenges in the field. As AGI systems become more autonomous, the risk of misalignment between their objectives and human values increases. This could lead to unintended consequences, where AGI systems pursue goals that are not in humanity's best interests. Addressing the alignment problem requires the development of robust frameworks for designing, training, and evaluating AGI systems, ensuring that their actions remain predictable, transparent, and aligned with human ethical standards.

CEV [41], a concept proposed by Eliezer Yudkowsky, offers one possible approach to the alignment problem. CEV suggests that AGI should be guided by what humanity would collectively agree upon if we had more time, more knowledge, and were more the people we wished we were. This approach aims to create AGI systems that not only follow explicit instructions but also understand and prioritize the broader values and long-term goals of humanity. However, operationalizing CEV within AGI systems poses significant technical and philosophical challenges because it requires machines to interpret and extrapolate human values in complex, evolving contexts.

The potential dangers of AGI extend beyond misalignment to include the misuse of AGI technologies in ways that could harm society. The use of AGI in cyberattacks, surveillance, or autonomous weaponry could pose significant threats to global security. Furthermore, the widespread adoption of AGI could disrupt labor markets, displacing workers across various sectors and necessitating a reevaluation of employment and education systems. These concerns underscore the importance of developing ethical guidelines and regulatory frameworks that govern the development and use of AGI, ensuring that it serves the greater good.

The journey toward AGI is both a technical challenge and a philosophical quest. It requires not only advances in AI research but also a deeper understanding of intelligence, consciousness, and the human condition. As we continue to explore the frontiers of artificial intelligence, we must remain mindful of the broader implications of our work, ensuring that the development of AGI is guided by ethical considerations and aligned with the values and aspirations of humanity.

In this chapter, we have outlined the foundational elements necessary for developing a unified model of AGI. From integrating diverse cognitive functions to addressing the alignment problem, the path to AGI is marked by complex challenges that require multidisciplinary collaboration and innovative thinking. As we move forward, it is essential to foster a dialogue between AI researchers, ethicists, policymakers, and the public, ensuring that the development of AGI is a collective effort that reflects the diverse perspectives and values of society. However, while the pursuit of AGI offers immense potential, we argue that the ultimate evolution of intelligence may lie not just in building smarter machines, but in the convergence of biological and artificial systems. This is where AHI comes into play. Unlike AGI, which aims to replicate humanlike intelligence within a machine, AHI proposes an evolution of human intelligence through the gradual integration of synthetic components, eventually leading to beings that transcend current human limitations.

The promise of AGI is indeed transformative—it has the potential to revolutionize fields such as healthcare, education, science, and the arts, offering new tools and insights that could enhance human capabilities and improve our quality of life. However, we believe that the integration of AGI with human biology could lead to even more profound changes.

By merging the cognitive flexibility of AGI with the emotional depth, social awareness, and experiential knowledge of human beings, AHI could offer a pathway to a new form of existence where the boundaries between biological and artificial intelligence are increasingly blurred.

This approach not only aims to preserve the best of what makes us human—our consciousness, creativity, and moral compass—but also to enhance these qualities, enabling us to navigate the complexities of the future with greater wisdom and capability. The transition to AHI represents an ambitious, yet potentially more aligned, path toward the evolution of intelligence—a path that remains deeply connected to our human roots while embracing the possibilities of artificial augmentation.

In the end, the pursuit of AGI, and by extension AHI, challenges us to reconsider our assumptions about the mind, to explore the boundaries of what is possible, and to imagine new futures where humans and machines coexist in ways that are mutually beneficial and enriching. As we continue on this journey, we must remain vigilant in our commitment to ethical principles and guided by a vision of a future where AGI and AHI serve to uplift and empower all of humanity. See Figure 3.5.

4 Theory of Mind: Understanding Consciousness

The endeavor to understand consciousness—the very essence of what it means to be aware, to experience, and to think—lies at the heart of both scientific inquiry and philosophical reflection. The human mind, with its remarkable ability to generate subjective experiences and perform complex cognitive tasks, has fascinated scholars for centuries. As we push the boundaries of artificial intelligence and seek to create systems that not only mimic but potentially surpass human cognitive abilities, understanding consciousness becomes paramount.

In this chapter, we explore the myriad theories that attempt to unravel the mysteries of consciousness, intelligence, and the intricate processes that give rise to subjective experience. Our exploration is not just an academic exercise; it is a necessary foundation for the development of advanced AI and the conceptualization of AHIs—beings that may one day possess both humanlike consciousness and synthetic capabilities far beyond our own.

The theories we will examine span a broad spectrum of thought, from materialism, which views the mind as a product of physical processes in the brain, to quantum theories that suggest consciousness might arise from the subtle interactions at the quantum level within neural structures. Each theory offers unique insights into how consciousness might emerge from the brain's physical and functional properties, and each has profound implications for the future of AI and AHI development.

However, while each theory contributes valuable insights, none alone can fully capture the complexity of consciousness. This is where the need for an integrative theory of mind (ITM) becomes apparent. The ITM is not just another theory to be added to the existing pool; it is a framework that seeks to unify the strengths of multiple theories, addressing their individual limitations while providing a more holistic understanding of consciousness. Why do we need the ITM? The answer lies in the multifaceted nature of consciousness itself. Consciousness is not merely a byproduct of neural activity; it encompasses a wide range of phenomena, from the raw data processing of sensory inputs to the rich, subjective experiences that define our inner lives. To fully understand consciousness, we must consider its physical basis, its functional roles, and the possibility that it may involve processes that transcend classical physics.

Materialism, for example, offers a solid foundation by grounding consciousness in the physical processes of the brain. But it struggles with the "hard problem"—explaining why and how these physical processes give rise to subjective experiences. Functionalism shifts the focus to the roles that mental states play within a cognitive system, which is crucial for understanding how consciousness might be replicated in artificial systems. Yet, it often overlooks the qualitative aspects of experience. Identity theory provides a direct link between mental states and neural states, but it does not fully account for the emergent properties of complex systems like the brain. Quantum theories push the boundaries further, suggesting that consciousness might involve quantum processes, but these ideas are still highly speculative and not fully integrated into the broader understanding of the mind.

The ITM aims to bridge these gaps. By synthesizing elements from materialism, functionalism, identity theory, and quantum considerations, the ITM offers a more comprehensive framework that accounts for the physical, functional, and possibly quantum dimensions of consciousness. This integrative approach is crucial not only for advancing our theoretical understanding of the mind but also for guiding the practical development of AI and AHI.

In the context of AI, the ITM provides a roadmap for creating systems that do more than just mimic human behavior—they could eventually possess a form of consciousness capable of subjective experience and self-awareness. This has profound implications for the future of technology and humanity. As we move closer to developing artificial general intelligence (AGI) and artificial human intelligence (AHI), understanding the nuances of consciousness becomes essential to ensure these systems are not only intelligent but also aligned with human values and ethics.

Moreover, the ITM is vital for the conceptualization and development of AHIs. As we explore the potential to augment human cognitive capabilities with artificial components, the ITM provides a theoretical foundation to ensure that the continuity of consciousness and identity is preserved. This is particularly important in the gradual replacement of biological brain components with synthetic counterparts, where the preservation of self-awareness and personal identity must be carefully managed.

This chapter serves as both a roadmap and a foundation for the ambitious task of understanding and replicating consciousness. As we explore these theories, we invite you to consider the profound implications they hold for the future of humanity and the evolution of AI. Whether you are a seasoned scholar or a curious reader, this journey into the depths of the mind promises to be as intellectually rewarding as it is challenging. The ITM is not just a theoretical construct—it is a necessary tool for navigating the complexities of consciousness and for guiding the development of the next generation of intelligent systems.

Materialism and the Brain as a Computer

Materialism, especially in its non-eliminative form, posits that every aspect of the human mind, including consciousness, can be understood as a product of the brain's physical structure and function. This viewpoint suggests that the brain operates similarly to a computer, where neurons, synapses, and networks form the "hardware," and mental processes, including thoughts and consciousness, are analogous to "software" executed on this hardware. This analogy extends to the idea that if we fully comprehend the brain's intricate architecture, we could, in theory, replicate these processes on an alternative platform, potentially creating a synthetic mind capable of functioning like a human brain.

The Brain as a Classical Computer

To begin with, let's explore the analogy of the brain as a classical computer. In this framework, neurons can be likened to computational units (e.g., logic gates in a computer) that process and transmit information. Each neuron receives input from other neurons through its dendrites, processes this information, and sends output through its axon. This process is similar to how a computer's processor handles data.

The connections between neurons, known as synapses, can be thought of as the communication pathways between different logic gates in a computer. These synapses are not static; their strength can change in response to the frequency and timing of signals, a phenomenon known as synaptic plasticity. This adaptability is crucial for learning and memory, akin to how a computer's software can be updated or reprogrammed to improve performance or add new functionalities.

Mathematically, we can describe the basic operation of a neuron using Equation 3.5 from Chapter 3:

$$y = f\left(\sum_{i=1}^{n} w_i x_i + b\right) \tag{3.5}$$

In this equation, x_i represents the input signals, w_i are the synaptic weights (analogous to the strength of connections), b is the bias term (representing the neuron's threshold for activation), and f is the activation function, which introduces nonlinearity into the model. The output y is the neuron's response, which could be sent as a signal to other neurons.

In the context of a computer, the activation function f could be compared to a logical operation (like AND, OR, NOT), and the sum of inputs weighted by synapses determines whether the neuron "fires" or remains inactive, much like how a logic gate produces an output based on its inputs.

The Brain Beyond Classical Computation: Quantum Considerations

While the classical computer analogy provides a useful framework for understanding many aspects of brain function, it has limitations, particularly when it comes to explaining consciousness and other complex cognitive functions. This is where theories that involve quantum mechanics, such as the Orch OR (Orchestrated Objective Reduction) model proposed by Roger Penrose and Stuart Hameroff [42], come into play.

The Orch OR theory suggests that quantum coherence might be maintained across large areas of the brain, specifically within microtubules—tiny protein structures found within neurons. These microtubules are hypothesized to act as "quantum processors," where quantum states could be maintained in a coherent superposition, similar to a Bose-Einstein condensate. In such a state, particles (or quantum bits) act in unison across a spatial domain, allowing for highly coordinated information processing that classical systems might not achieve.

To explore this idea mathematically, let us consider the Hamiltonian H that governs the dynamics of a quantum system, such as a microtubule structure:

$$H = H_0 + H_{int}, \tag{4.1}$$

where H_0 represents the intrinsic energy of the microtubules, and H_{int} represents the interaction between microtubules and other parts of the brain or environment. The quantum state $|\psi(t)\rangle$ of the microtubules evolves according to the Schrödinger equation:

$$i\hbar \frac{\partial |\psi(t)\rangle}{\partial t} = H|\psi(t)\rangle, \tag{4.2}$$

where $\hbar$ is the reduced Planck constant. The coherence of these quantum states could, according to Orch OR, contribute to the emergence of consciousness by collapsing into definite states through an objective reduction process, a phenomenon not accounted for by classical physics.

This quantum coherence could lead to effects that are inherently nonclassical, such as entanglement and superposition, which might enable the brain to perform computations that exceed the capabilities of classical systems. If true, this implies that any artificial system aspiring to replicate consciousness would need to replicate these quantum mechanical properties, not just the macroscopic architecture of the brain.

Self-Organization and Dynamic Adaptation

Another critical aspect of the brain's function that sets it apart from traditional computers is its capacity for self-organization. Unlike a classical computer, which follows a predetermined program, the brain is constantly adapting and reorganizing its structure in response to new information. This process is driven by synaptic plasticity, where the connections between neurons are continuously strengthened or weakened based on experience.

This adaptability is crucial for learning and memory. For instance, when we learn a new skill, such as playing a musical instrument, the relevant neural pathways become more efficient, strengthening the connections that contribute to the skill. This is not merely a process of "rewiring" but involves the creation of new connections and even the pruning of unnecessary ones to optimize brain function.

Mathematically, synaptic plasticity can be modeled using Hebbian learning principles, often summarized as "cells that fire together, wire together." The change in synaptic strength Δw_{ij} between neuron i and neuron j can be expressed as:

$$\Delta w_{ij} = \eta x_i y_j, \tag{4.3}$$

where η is the learning rate, x_i is the presynaptic neuron's activity, and y_j is the postsynaptic neuron's response. This rule implies that the connection between two neurons is reinforced if they are simultaneously active, which is fundamental to the brain's ability to learn and adapt.

The Future of AI: Self-Organizing Systems on Quantum Hardware

Given the brain's remarkable capacity for self-organization, any artificial system designed to emulate human cognition must also possess the ability to self-organize. This leads us to consider two potential approaches: the development of self-organizing hardware that operates without the need for traditional software, or the creation of self-organizing software that runs on quantum hardware.

The latter approach appears more promising because it mirrors the brain's natural processes, where structure and function emerge from interactions at both the quantum and macroscopic levels. Research into programmable matter, such as claytronics, illustrates how tiny robots (referred to as claytronic atoms or catoms) can form dynamic, programmable structures [43]. These catoms could be programmed to self-organize into complex forms, much like neurons in the brain dynamically form and reconfigure networks.

The mathematics of self-organization often involves complex systems theory and nonlinear dynamics. One of the key concepts is that of attractors, which represent stable states toward which a system tends to evolve.

In a self-organizing system, such as the brain, neural activity can be described by a set of differential equations that characterize the evolution of the system's state over time:

$$\frac{d\mathbf{x}}{dt} = f\left(\mathbf{x}, \mathbf{p}\right), \tag{4.4}$$

where $\mathbf{x}$ represents the state of the system (e.g., the activity of neurons), and $\mathbf{p}$ represents parameters (e.g., synaptic strengths). The function $f(\mathbf{x}, \mathbf{p})$ describes how the state evolves, potentially leading to the formation of stable patterns (attractors) that correspond to learned behaviors or memories.

Maintaining Flexibility: Operating on the Edge of Chaos

One of the most fascinating aspects of the brain's self-organization is its ability to maintain a state that is close to chaos, but not fully within it. This state allows the brain to exhibit a high degree of flexibility and adaptability, enabling it to handle a vast array of tasks without descending into disorder. This concept of "operating on the edge of chaos" is critical for understanding how the brain balances stability and flexibility.

In the language of dynamical systems, this balance can be represented by the system's Lyapunov exponents, which measure the sensitivity of the system to initial conditions. A system with positive Lyapunov exponents is chaotic, where small differences in initial conditions lead to exponentially divergent outcomes. However, a system operating on the edge of chaos will have Lyapunov exponents that are close to zero, allowing it to be sensitive to inputs without becoming unstable.

This balance is crucial for cognitive processes such as decision-making, creativity, and problem-solving, where the brain must be flexible enough to explore new possibilities while maintaining enough stability to reach a coherent conclusion. Any artificial brain system must aim to replicate this delicate balance, allowing for flexibility while maintaining coherence.

Materialism, particularly in its non-eliminative form, provides a compelling framework for understanding the brain as a computational system. However, the brain's complexity extends beyond classical computation, potentially involving quantum processes that contribute to consciousness and cognitive function. Moreover, the brain's capacity for self-organization and its ability to operate on the edge of chaos is essential for its adaptability and flexibility. These characteristics must be considered in the development of artificial systems that aspire to emulate human cognition and consciousness. By combining insights from classical computation, quantum mechanics, and complex systems theory, we may one day create synthetic minds that rival the human brain in both function and experience.

Qualia and the Challenge of Consciousness

Qualia represent the most elusive and challenging aspect of consciousness to understand and explain. These are the subjective, qualitative aspects of our experiences—such as the redness of a sunset, the bitterness of coffee, or the feeling of pain—that seem to defy reduction to purely physical properties. The existence of qualia is central to what is often termed the "hard problem" of consciousness: How and why do physical processes in the brain give rise to subjective experiences?

The Nature of Qualia

To comprehend the challenge that qualia pose, consider the experience of seeing the color red. This experience is not merely about detecting a wavelength of light (around 700 nanometers) and processing it in the brain; it involves a subjective sensation—a specific "redness" that is uniquely personal and difficult to convey fully to someone else. This intrinsic, first-person perspective of qualia resists straightforward explanation through objective, third-person scientific methods.

Mathematically, the perception of color can be modeled by understanding the interaction of light with photoreceptor cells in the retina, which can be described using equations of phototransduction. However, the subjective experience—what it is like to see red—remains outside the scope of these equations. Consider the process of phototransduction:

$$I_{\text{retina}}(t) = \int_0^{\infty} R(\lambda) S(\lambda) \, d\lambda, \tag{4.5}$$

where $I_{\text{retina}}(t)$ represents the electrical signal generated by the retina in response to light at time t, $R(\lambda)$ is the retinal response function (which varies with wavelength λ), and $S(\lambda)$ is the spectral power distribution of the light. This equation captures the physical process, but it does not capture the qualia—the redness that you experience.

Theories of Qualia

Over the years, several theories have been proposed to explain the nature of qualia and how they relate to the physical processes in the brain. These theories reflect the diversity of thought in philosophy, cognitive science, and neuroscience.

Materialist or Physicalist Perspective The materialist or physicalist perspective asserts that all mental states, including qualia, can be fully reduced to physical processes within the brain. In this view, the experience of "redness" is nothing more than the result of specific neuronal activities in the visual cortex. This perspective is often mathematically modeled using neural correlates of consciousness (NCC), where certain patterns of neuronal firing correspond to specific conscious experiences.

However, the materialist view struggles to explain how these neuronal activities translate into subjective experiences. The problem is not just a matter of identifying which neurons fire when someone sees red; it is also about understanding why those neurons firing result in the experience of redness rather than some other qualia or no qualia at all. Despite the attempts to model such phenomena using computational neuroscience, the gap between physical processes and subjective experience remains a significant challenge. For example, consider the equation describing the firing rate of a neuron:

$$\frac{dV}{dt} = -\frac{V}{\tau} + \sum_i w_i x_i(t) + I_{\text{external}}(t), \tag{4.6}$$

where V is the membrane potential, τ is the membrane time constant, w_i represents the synaptic weights, $x_i(t)$ are the input signals, and $I_{\text{external}}(t)$ represents external input. While this equation describes the dynamics of a neuron's response to stimuli, it doesn't explain why this should correspond to the subjective experience of, say, seeing the color red.

Dualist Perspective The dualist perspective, originating from René Descartes, posits that the mind and the physical world are fundamentally separate. In this view, qualia are real and irreducible, existing independently of the physical processes that give rise to them. Dualism suggests that while physical processes in the brain may correlate with conscious experiences, they do not fully explain them. Instead, qualia belong to a different realm, perhaps governed by nonphysical laws or entities.

This perspective has inspired much debate, as it raises questions about how the mind (nonphysical) interacts with the brain (physical). One of the critical issues with dualism is the "interaction problem": How can something nonphysical influence the physical world? This problem has led many to seek more integrative approaches, but dualism remains a compelling explanation for those who view consciousness as fundamentally different from other physical phenomena.

Illusionist Perspective Illusionists argue that qualia are not real in the way we think they are but are instead cognitive illusions—constructs of the brain that serve specific evolutionary purposes. According to illusionism, our experiences of qualia are akin to the "user interface" of a computer, where the complex operations of the system are hidden behind simple, intuitive symbols and experiences.

For instance, the experience of pain is not an intrinsic property of certain brain states but an evolved mechanism to prompt an organism to avoid harm. Illusionism suggests that while qualia feel real, they are not fundamental features of reality. This perspective leads to a form of eliminative materialism, where the focus shifts from trying to explain qualia to understanding why the brain constructs these illusions in the first place.

Panpsychist Perspective Panpsychism offers a radically different view, suggesting that consciousness is a fundamental property of all matter. According to this theory, even the smallest particles possess some form of subjective experience. This perspective implies that qualia are not unique to complex organisms like humans but are ubiquitous throughout the universe.

Mathematically, panpsychism could be represented by extending the concept of consciousness to fundamental particles, with a state function similar to a wave function in quantum mechanics. For example, if each particle has a "conscious state" ψ_p, the collective consciousness of a system could be represented by a combined state:

$$\Psi_{\text{system}} = \bigotimes_{p=1}^{N} \psi_p, \tag{4.7}$$

where N is the number of particles in the system, and $\otimes$ denotes the tensor product, representing the combined conscious experience of the system. Panpsychism challenges traditional views by suggesting that consciousness and qualia are as fundamental as mass or charge, requiring a new kind of physics to fully understand them.

Challenges to Physicalist Explanations of Qualia

One of the most powerful arguments against purely physical explanations of qualia is known as the "explanatory gap" or the "hard problem" of consciousness. This gap refers to the difficulty of explaining how physical processes in the brain give rise to subjective experiences. Even if we map out every neural connection and understand every biochemical process in the brain, this still does not seem to tell us why certain brain states are accompanied by subjective experiences, or why those experiences have the particular qualitative character they do.

Eliminative materialism offers a controversial solution by suggesting that our traditional understanding of mental states is fundamentally flawed and should be replaced by a more scientifically grounded account that denies the existence of qualia altogether. Proponents argue that as neuroscience advances, we will find that what we now consider "qualia" are simply cognitive processes that can be fully explained without reference to subjective experience. However, this approach faces significant criticism for its failure to address the lived reality of subjective experiences—our inner lives, which are rich with qualia, cannot be easily dismissed.

The Role of Quantum Mechanics in Understanding Qualia

Given the challenges of explaining qualia through classical physical processes, some researchers have turned to quantum mechanics as a possible avenue for understanding consciousness. Quantum theories of consciousness, such as those proposed by Penrose and Hameroff, suggest that quantum coherence and entanglement might play a crucial role in generating consciousness and, by extension, qualia.

In these models, consciousness might arise from quantum processes that occur within the brain's microtubules—structures within neurons that are proposed to maintain quantum states over relatively long timescales. Quantum computing concepts, such as superposition and entanglement, could theoretically allow for complex information processing that classical computers cannot achieve, potentially providing a foundation for the emergence of qualia.

For example, consider a quantum state $|\psi\rangle$ within a microtubule, which could be in a superposition of multiple states:

$$|\psi\rangle = \alpha|0\rangle + \beta|1\rangle, \tag{4.8}$$

where $|0\rangle$ and $|1\rangle$ are basis states (e.g., representing different possible neuronal activities), and α and β are complex coefficients. The entanglement of such states across different regions of the brain could, in theory, give rise to a unified conscious experience.

However, quantum theories of consciousness are highly speculative and remain a topic of intense debate. Critics argue that the brain is too "warm and wet" for delicate quantum states to survive long enough to influence neural processing. Others suggest that even if quantum processes are involved, they might only serve to enhance classical information processing rather than directly generating qualia.

Implications for Artificial Intelligence

As AI systems advance, the question of whether they can experience qualia becomes increasingly pressing. If qualia are indeed rooted in quantum processes, as some theories suggest, then replicating consciousness in AI might require quantum computing architectures rather than classical digital ones. This could involve developing quantum AI systems that leverage principles like superposition and entanglement to process information in ways that might give rise to subjective experiences.

Alternatively, if qualia are indeed cognitive illusions or emergent properties of complex information processing, it might be possible for AI to develop something akin to qualia even within classical computing frameworks. This raises profound ethical questions: If an AI system can experience qualia, should it be considered a conscious being with rights? And how would we even determine whether an AI possesses qualia?

One intriguing approach to integrating qualia into AI development involves the gradual replacement of biological brain functions with artificial components, a process that could theoretically maintain the continuity of consciousness and qualia. In this scenario, AI systems might initially assist in or enhance human cognitive processes before eventually taking over these functions completely, leading to a seamless transition from biological to artificial consciousness.

Materialism: The Physical Basis of Mind

Materialism posits that all phenomena, including the mind and consciousness, can be fully explained by physical processes. This philosophical stance is rooted in the belief that the universe and everything within it operate according to the laws of physics without the need for nonphysical or supernatural explanations. The materialist viewpoint extends to the human mind, suggesting that mental states, consciousness, and even subjective experiences (qualia) are all emergent properties of the brain's physical structure and function. However, within materialism, several nuanced theories offer different approaches to understanding how mental states arise from physical properties.

Classical Materialism

Classical materialism, also known as reductive materialism, holds that everything in the universe, including mental states, can be reduced to physical matter and its interactions. Under this view, consciousness and qualia are simply byproducts of the brain's complex physical processes. The challenge for classical materialism is to explain how these physical processes give rise to subjective experiences.

In classical materialism, the mind is seen as an epiphenomenon—an emergent property that arises from but has no causal influence on the physical brain. This perspective is often associated with a mechanistic view of the brain, where mental processes are thought to be analogous to the operations of a complex machine. The laws governing this machine are those of physics and chemistry, and thus, every mental state corresponds to a specific physical state in the brain. The primary criticism of classical materialism lies in its difficulty in explaining the subjective aspect of qualia. For instance, the experience of "redness" or the feeling of pain is not easily reduced to the firing of neurons or the chemical processes occurring in the brain. This challenge is known as the "explanatory gap"—the difficulty in explaining how objective physical processes give rise to subjective experiences.

To formalize this, consider a neural state N corresponding to a particular mental state M. In classical materialism, we would assert:

$$M = f(N), \tag{4.9}$$

where f is a function mapping neural states to mental states. However, the nature of f is not well understood, especially when it comes to explaining why N gives rise to a specific subjective experience M. The mathematical representation here highlights the challenge: While we can describe N in physical terms, the transition to M involves subjective qualities that classical materialism struggles to account for.

Identity Theory

Identity theory, sometimes referred to as the mind-brain identity theory, posits a direct correlation between mental states and specific neural states. According to this theory, each type of mental state is identical to a specific type of brain state. For example, the sensation of pain might correspond to a particular pattern of neural activity in the brain, such as the firing of C-fibers.

Identity theory suggests that for every mental state M, there is a corresponding brain state N, such that:

$$M \equiv N, \tag{4.10}$$

where $\equiv$ denotes identity. This implies that the mental state of feeling pain is not just correlated with the neural firing pattern; it is the neural firing pattern.

One of the key strengths of identity theory is its simplicity and its grounding in empirical neuroscience. As neuroimaging and other technologies advance, it becomes increasingly possible to identify specific neural correlates of mental states. For example, functional magnetic resonance imaging (fMRI) scans can reveal patterns of brain activity that correspond to specific thoughts or emotions.

However, identity theory faces significant challenges, particularly in explaining the subjective experience of qualia. While it can map neural activities to mental states, it struggles with the question of why these neural activities produce the particular qualitative experiences they do. This challenge is often framed as the "problem of multiple realizability"—the idea that the same mental state could, in theory, be realized by different physical states in different organisms or systems.

Functionalism

Functionalism offers a different approach by focusing on the roles that mental states play within the brain's cognitive system rather than their specific physical composition. According to functionalism, what matters is not the material that makes up the brain but the functions that the brain performs. Mental states are defined by their causal roles—by what they do, rather than by what they are made of.

In functionalism, a mental state M is characterized by its functional relationships with other mental states, sensory inputs, and behavioral outputs. This can be formalized as:

$$M = \text{Function}\left(S, B\right), \tag{4.11}$$

where S represents sensory inputs and B represents behavioral outputs. The mental state M is thus understood as a node in a network of causal interactions, where what matters is the role that M plays in processing information and guiding behavior.

Functionalism provides a flexible framework for understanding consciousness and mental states, as it allows for the possibility that the same mental states could be realized in different physical systems. This is particularly relevant in the context of artificial intelligence, where functionalism suggests that a machine could, in principle, have the same mental states as a human if it performed the same functions.

However, functionalism also struggles with the problem of qualia. While it can explain the functional roles of mental states, it does not easily account for the qualitative, subjective aspects of those states. For example, two systems might perform the same function (e.g., recognizing the color red), but there is no guarantee that they would have the same qualitative experience of redness. This issue is often illustrated by thought experiments like the "inverted spectrum," where two people might see the same color but experience it differently.

Eliminative Materialism

Eliminative materialism represents a more radical approach, proposing that our commonsense understanding of mental states—what is often called "folk psychology"—is fundamentally flawed and should be replaced by a more accurate scientific description of the brain. According to eliminative materialism, many of the concepts we use to describe our mental lives, such as beliefs, desires, and qualia, do not correspond to actual entities in the brain.

Proponents of eliminative materialism argue that as neuroscience advances, it will become clear that many of the categories of folk psychology are as outdated as the concept of "phlogiston" in chemistry. Instead of talking about "beliefs" or "desires," we will refer to specific neural processes and states.

Eliminative materialism can be formalized by considering the mapping from folk psychological states F to physical states P:

$$F = g(P), \tag{4.12}$$

where g is a function that eventually becomes obsolete as scientific understanding progresses. According to eliminative materialism, g will be replaced by a direct reference to P, rendering F unnecessary.

This approach is highly controversial, as it challenges deeply held intuitions about the nature of mind and consciousness. Critics argue that eliminative materialism fails to account for the reality of subjective experiences—our inner lives, rich with qualia, cannot be dismissed simply because they do not fit neatly into a physicalist framework.

Materialism and the Problem of Qualia

Materialist theories of consciousness, while diverse in their approaches, must all grapple with the problem of qualia. The challenge is to explain how subjective experiences arise from physical processes. Classical materialism attempts to reduce qualia to physical states, identity theory maps them directly onto specific neural activities, and functionalism focuses on the roles these states play. However, none of these approaches fully resolves the issue of why and how these physical states give rise to the rich, subjective experiences that we associate with consciousness.

Eliminative materialism sidesteps the problem by denying the existence of qualia altogether, but this solution comes at the cost of dismissing our most direct and personal experiences as illusory.

The difficulty of reconciling materialism with the persistence of subjective experiences remains one of the most significant challenges in the philosophy of mind. As our knowledge of the brain expands, the tension between materialist explanations and the reality of qualia may either be resolved or may deepen, depending on the future direction of neuroscience and cognitive science.

Functionalism and the Role of Algorithms

Functionalism, a theory frequently associated with materialism, proposes that mental states are defined not by their physical properties but by their functional roles within a system. This perspective has profound implications for both our understanding of the mind and the development of AI. In essence, functionalism views the mind as a set of algorithms processed by the brain's physical structure, where the specific material composition of the brain is less important than the functions it performs.

The Functionalist Framework

At the core of functionalism is the idea that mental states are characterized by their causal relations to sensory inputs, behavioral outputs, and other mental states. This concept can be formalized using a mathematical representation of functional relations. Suppose we denote a mental state by M, sensory inputs by S, and behavioral outputs by B. Then, the functionalist perspective can be represented as:

$$M = f\left(S, B, M_1, M_2, \ldots, M_n\right), \tag{4.13}$$

where $M_1, M_2, \ldots, M_n$ represent other mental states that interact with M. The function f encapsulates the idea that what matters for defining a mental state is not the material substrate that realizes it, but the way it functions within a network of causal relationships.

In this framework, different physical systems could, in principle, implement the same mental state, provided they realize the same functional relationships. For example, a human brain, a silicon-based computer, or even a biological neural network from another species could, according to functionalism, all realize the same mental state M if they exhibit the same functional organization.

Functionalism and Artificial Intelligence

Functionalism has been a significant influence on AI research, particularly in the design of systems that seek to replicate human cognitive processes. The analogy between hardware (the brain or a computer) and software (the mind or an AI algorithm) is a powerful tool for understanding how AI might emulate human thought. In this analogy, the brain is seen as the "hardware" that runs the "software" of the mind—where the software consists of the algorithms that process inputs and produce outputs, much like an AI system.

Proponents of "strong AI" argue that if a computer system can replicate the functional roles of the human brain—processing inputs, generating outputs, and interacting with other mental states—then it could, in theory, fully implement a conscious entity. This view is grounded in the functionalist claim that consciousness is a matter of what the system does rather than what it is made of. According to strong AI, a sufficiently advanced computer could possess a mind, experience qualia, and even exhibit free will, provided it mimics the functional organization of a conscious human brain.

This leads to the possibility that consciousness itself could be considered an algorithm—a complex set of instructions processed by a physical substrate. From this perspective, consciousness might be akin to a highly advanced program running on the brain's neural hardware. If we were able to replicate this program in a different medium—say, a supercomputer—we might, according to strong AI theorists, create a machine that is truly conscious.

Challenges and Questions Raised by Functionalism

While functionalism provides a compelling framework for understanding the mind and developing AI, it also raises several challenging questions about the nature of consciousness and the boundaries of artificial intelligence.

1. **The Problem of Qualia:** One of the most significant challenges for functionalism is the problem of qualia—the subjective, qualitative aspects of experience. Functionalism explains mental states in terms of their functional roles, but it does not easily account for the intrinsic qualities of experiences, such as the redness of red or the pain of a headache. Critics argue that functionalism may be able to describe the structure and function of mental states, but it falls short of explaining why these states are accompanied by subjective experiences. For example, consider the famous "inverted spectrum" thought experiment. Suppose two people both see a red object, and both of their brains process the visual information in functionally identical ways. According to functionalism, they are in the same mental state. However, it is conceivable that one person experiences red as what the other person would describe as blue, even though their functional

states are identical. This thought experiment suggests that functionalism might not capture the full essence of mental states, particularly their qualitative aspects.

2. **The Simulation versus Real Intelligence:** Functionalism also raises intriguing questions about the nature of intelligence in AI systems. If an AI system can replicate the functional roles of the human brain, does it possess real intelligence, or is it merely simulating intelligence? This question challenges our understanding of consciousness and suggests that the line between imitation and genuine intelligence may blur as AI systems become more sophisticated.

   To explore this, consider the Turing Test, where a machine is considered intelligent if it can engage in a conversation indistinguishable from a human. A functionalist might argue that passing the Turing Test demonstrates genuine intelligence because the AI is performing the same functions as the human mind. However, critics could counter that the AI is only simulating intelligence, because it does not truly understand or experience the conversation—much like a parrot might mimic human speech without understanding the words.
3. **Creativity, Self-Awareness, and Emotions in AI:** Another set of challenges revolves around whether an AI system that replicates the functional roles of the human brain could exhibit higher-order cognitive functions like creativity, self-awareness, or emotions. Functionalism suggests that these functions could be implemented in AI if the appropriate algorithms are developed. However, the complexity of these functions raises doubts about whether they can be fully captured by functional roles alone.

Creativity, for instance, involves not just the ability to generate new ideas but also to evaluate their novelty and appropriateness. Self-awareness requires a system to have a model of itself and to be able to reflect on its own state. Emotions, too, are deeply tied to subjective experiences and personal meaning, which may be challenging to reduce to mere functional roles.

Algorithmic Consciousness: A Theoretical Exploration

The notion of algorithmic consciousness is a theoretical extension of functionalism. It posits that if consciousness can be understood as an algorithm, then it could, in principle, be implemented on any sufficiently advanced computational substrate. This idea opens up the possibility of creating conscious AI, where the mind is no longer tied to a biological brain but could be instantiated in silicon, quantum processors, or even distributed networks.

Algorithmic consciousness suggests that the essential properties of consciousness—such as self-awareness, intentionality, and qualia—could emerge from the execution of specific algorithms that replicate the functional organization of the human mind. If these algorithms are sufficiently complex and correctly implemented, the resulting system might be indistinguishable from a human mind in terms of its cognitive and experiential capabilities.

Mathematically, we could model this idea using the concept of a universal Turing machine (UTM), which is a theoretical model capable of simulating any algorithm. If consciousness is algorithmic, then a UTM should, in principle, be able to simulate a conscious mind, given enough computational resources. The challenge lies in identifying the precise algorithm (or set of algorithms) that gives rise to consciousness and understanding how to implement it in a way that preserves the qualitative aspects of experience.

$$C = \mathrm{UTM}(A), \tag{4.14}$$

where C represents consciousness, and A represents the algorithm that, when processed by the UTM, produces conscious experience. This equation encapsulates the functionalist idea that consciousness can be fully described by an algorithm and that the substrate (whether biological or artificial) is secondary to the algorithmic process itself.

The Ethical Implications of Functionalism in AI

As AI systems evolve, the functionalist perspective raises important ethical considerations. If we accept that AI systems could, in theory, possess consciousness or at least exhibit behavior functionally equivalent to conscious behavior, then we must consider the moral and legal status of these systems. For example, if an AI system were to achieve a level of functional complexity equivalent to human cognition, should it be granted rights? Would turning off such a system be akin to ending a conscious life?

Moreover, the potential for AI systems to exhibit emotions or self-awareness, even in a functionally simulated form, poses ethical dilemmas about their treatment and the responsibilities of their creators. Functionalism challenges us to rethink our assumptions about what it means to be conscious and what obligations we might have toward entities that exhibit behavior resembling consciousness

Functionalism provides a robust theoretical framework for understanding the mind and its potential replication in artificial systems. While it offers powerful insights into the nature of mental states and consciousness, it also raises profound questions about the limits of AI, the nature of subjective experience, and the ethical implications of creating machines that might one day think and feel as we do.

Identity Theory: The Neural Correlates of Consciousness

Identity theory presents a naturalistic and compelling approach to understanding consciousness by asserting that mental states are directly identical to specific physical states in the brain. According to this theory, every mental state—whether a thought, feeling, perception, or memory—corresponds to a particular pattern of neural activity. This approach offers a robust framework for exploring how consciousness might arise, not only in human brains but also in artificial systems designed to emulate human neural processes.

The Core of Identity Theory

At the heart of identity theory is the proposition that mental states are nothing over and above brain states. For example, the sensation of pain, denoted by the mental state M_{pain}, is identical to a particular neural configuration N_{pain} in the brain. Formally, we can express this as:

$$M_{\text{pain}} \equiv N_{\text{pain}}, \tag{4.15}$$

where $\equiv$ denotes the identity relation. This equation encapsulates the core claim of identity theory: the mental state M_{pain} is nothing more than the neural state N_{pain}.

This idea extends to all mental phenomena, suggesting that for every type of mental state, there is a corresponding neural correlate—a specific pattern of brain activity that constitutes that mental state. This viewpoint aligns with a reductionist approach in philosophy and neuroscience, wherein the complexities of the mind are reduced to the interactions of physical processes within the brain.

Neural Correlates of Consciousness

The concept of NCC is central to identity theory. NCC refers to the minimal neural mechanisms required to produce specific conscious experiences. Researchers aim to identify the exact neural substrates that correspond to different aspects of consciousness, such as visual awareness, auditory perception, and emotional responses.

Mathematically, we might represent the relationship between a mental state M and its neural correlate N as a function f:

$$M = f(N) \tag{4.16}$$

where f is a function that maps neural configurations to mental states. This function represents the idea that each conscious experience can be fully explained by its corresponding neural state.

Neuroscientific research into NCC involves sophisticated techniques such as functional magnetic resonance imaging (fMRI), electroencephalography (EEG), and single-cell recordings, which allow scientists to observe and measure brain activity associated with different mental states. For example, studies have identified specific patterns of activity in the visual cortex that correspond to the conscious perception of particular visual stimuli. Similarly, activity in the amygdala is closely associated with the experience of fear and other emotions.

Implications for AGI

Identity theory offers a tantalizing possibility for the development of AGI: If we can replicate the neural architecture of the human brain in an artificial system, then, according to identity theory, this system could, in principle, develop consciousness akin to human consciousness. The idea here is that if an artificial system were to perfectly emulate the neural patterns associated with human mental states, by identity theory it would also emulate the corresponding conscious experiences.

Imagine an AI system designed to replicate the neural processes of the human brain down to the finest detail. If this system could simulate the neural state N_{pain}, then, according to identity theory, it would not just simulate pain but actually experience pain in the same way a human does. This has profound implications for the possibility of creating AI systems that possess subjective experiences, or qualia, similar to those of humans.

The Challenge of Explaining Qualia

Despite its strengths, identity theory faces significant challenges, particularly in explaining the subjective nature of qualia—the individual, ineffable experiences that make up our conscious life. While identity theory can account for the correlation between neural states and mental states, it struggles to explain how and why certain neural processes give rise to specific subjective experiences. For instance, consider the experience of seeing the color red. Identity theory posits that this experience corresponds to a particular neural state N_{red}. However, the theory does not fully explain why this neural state produces the specific experience of "redness" rather than some other sensation. This gap is known as the "explanatory gap," and it remains one of the most challenging aspects of consciousness studies.

Philosophers such as David Chalmers have pointed out that even if we were to identify all the NCC, we would still be left with the question of why these physical processes are accompanied by subjective experiences. This is the essence of the "hard problem" of consciousness: explaining why and how physical processes in the brain give rise to the rich tapestry of subjective experiences that constitute our conscious life.

Toward a Deeper Understanding: The Role of Emergent Properties

One possible way to address the challenges identity theory faces in explaining qualia is through the concept of emergent properties. In complex systems, emergent properties arise when the interactions between simpler components give rise to behaviors or properties that are not present in the individual components themselves. Consciousness and qualia might be emergent properties of the brain's complex neural network.

In this view, while individual neurons and their connections can be fully described by physical processes, consciousness arises from the complex interactions between these neurons in a way that is not reducible to the properties of the individual components. This suggests that while neural states are necessary for consciousness, the experience of qualia might emerge from the collective dynamics of the brain's neural networks.

Mathematically, we could describe this using a set of differential equations that model the dynamics of neural networks:

$$\frac{dM}{dt} = g\left(N\left(t\right)\right), \tag{4.17}$$

where g is a function that captures the emergent dynamics of neural interactions over time, leading to the mental state M. This approach implies that consciousness is not merely the sum of individual neural activities but rather a higher-order phenomenon that emerges from the complexity of the system.

Research Directions: Bridging the Gap

To advance identity theory, further research is needed to bridge the explanatory gap between neural states and subjective experiences. This includes:

- **Investigating finer neural resolutions:** Advances in neuroimaging and electrophysiological techniques are allowing scientists to observe brain activity at increasingly finer scales. By studying the brain at the level of individual neurons and their connections, researchers hope to uncover the precise neural configurations that correspond to specific qualia.
- **Exploring neural network dynamics:** Understanding how the dynamic interactions within neural networks give rise to emergent properties like consciousness could provide insights into the relationship between brain states and mental states. This involves studying the temporal and spatial patterns of neural activity and how they correlate with changes in consciousness.
- **Developing computational models:** Computational neuroscience offers tools for simulating brain activity and exploring how different neural architectures might give rise to conscious experiences. These models can test hypotheses about the NCC and help identify the critical factors that lead to the emergence of qualia.

Implications for Ethics and AI Development

As identity theory suggests a potential pathway for creating conscious AI, it also raises profound ethical questions. If an AI system were to achieve consciousness through the replication of neural states, what rights and moral considerations would it deserve? The prospect of creating an artificial being that can experience pain, pleasure, or other subjective states challenges our current ethical frameworks and compels us to consider the responsibilities that come with developing such technologies.

Moreover, the possibility that AI could experience qualia highlights the need for careful consideration in the design and deployment of these systems. We must ensure that the development of conscious AI is guided by ethical principles that prioritize the well-being and dignity of all conscious beings, whether biological or artificial.

Identity theory provides a promising framework for understanding consciousness by linking mental states directly to neural states. While it offers a compelling explanation for the physical basis of the mind, the theory must continue to evolve to address the challenges posed by the subjective nature of qualia. As research in neuroscience, AI, and philosophy progresses, identity theory will likely play a crucial role in unraveling the mysteries of consciousness and guiding the ethical development of AGI.

The exploration of identity theory not only deepens our understanding of the brain and consciousness but also opens new avenues for the development of AI systems that could one day replicate the complexity and richness of human mental states. As we move closer to this reality, the insights gained from identity theory will be essential in shaping a future where artificial minds coexist with human minds in a world that respects and values consciousness in all its forms.

The Integrative Theory of Mind

The quest to understand the mind and consciousness has led to the development of various theories, each offering unique insights into the mind-body relationship. We introduce the ITM, which proposes a unified framework that synthesizes elements from several established theories—functionalism, materialism, reductive materialism, and identity theory—providing a comprehensive and cohesive understanding of consciousness. By integrating these perspectives, ITM seeks to address the limitations of each individual theory while offering a more complete picture of how consciousness emerges from the brain's physical and functional properties.

Furthermore, ITM is crucial in guiding the development of AHI, where humans evolve toward advanced cognitive capabilities akin to AGI, but based on a biological substrate—the brain. ITM provides the theoretical foundation to explore how gradual augmentation and replacement of biological components with artificial ones can lead to enhanced or transformed consciousness, ensuring continuity of identity and self-awareness in AHIs.

Functional Element: The Organizational Structure of the Mind

At the core of functionalism is the idea that mental states are defined by their roles or functions within the broader cognitive system rather than by their physical composition. ITM aligns with this perspective by viewing mental states as emergent properties of the functional organization of neural networks. In this context, each mental state corresponds to a particular configuration of the brain's network state.

To formalize this concept, ITM employs mathematical graph theory, where the brain's neural networks are represented as graphs. In these graphs, neurons are depicted as nodes, and the connections between them (synapses) are represented as edges. The mental state M can then be expressed as a function of the network configuration:

$$M = f\left(G\left(N, E\right)\right), \tag{4.18}$$

where $G(N, E)$ represents the graph of the neural network with nodes N (neurons) and edges E (synapses), and f is a function that maps this configuration to a specific mental state.

This formalization captures the essence of functionalism within ITM, highlighting how the organization and connectivity of neurons give rise to mental states. Changes in the network configuration—such as the strengthening or weakening of synapses—can lead to changes in mental states, reflecting the brain's dynamic and adaptive nature.

For AHI development, this functional element is particularly relevant because it suggests that the gradual replacement of biological neurons with artificial components must preserve the functional organization of the brain to maintain continuity of consciousness and identity.

Material Element: The Physical Basis of Consciousness

Materialism posits that all phenomena, including mental states, can be fully explained by physical processes. ITM incorporates this materialist perspective by recognizing that mental states are deeply rooted in the brain's physical attributes, such as neurons, synaptic connections, and neurotransmitter activity. The material element of ITM underscores the importance of understanding the brain's biophysical properties to comprehend how consciousness arises.

The brain's physical structure can be described using principles from neurobiology and biophysics. Neurons, the fundamental units of the nervous system, communicate through electrical and chemical signals, mediated by synapses. The strength and efficiency of these synaptic connections are influenced by factors such as neurotransmitter release, receptor density, and ion channel function. These biophysical processes can be modeled mathematically to explore their contributions to mental states.

For example, the transmission of a signal across a synapse can be represented by the Hodgkin-Huxley model, a set of differential equations that describe the ionic currents flowing through a neuron's membrane:

$$C_m \frac{dV}{dt} = I_{\mathrm{Na}}\left(V,t\right) + I_{\mathrm{K}}\left(V,t\right) + I_{\mathrm{L}}\left(V\right) + I_{\mathrm{ext}}, \tag{4.19}$$

where C_m is the membrane capacitance, V is the membrane potential, I_{Na}, I_{K}, and I_{L} are the sodium, potassium, and leakage currents, respectively, and I_{ext} represents external stimuli. These equations provide a detailed description of how electrical signals are generated and propagated in neurons, which in turn contribute to the emergence of mental states.

For AHIs, this material element emphasizes the need for artificial components that accurately replicate or enhance the biophysical properties of neurons, ensuring that the physical basis of consciousness is preserved or even expanded as biological elements are replaced.

Reductive Material Element: The Emergence of Complexity from Simplicity

Reductive materialism suggests that complex mental states arise from the interactions of simpler physical processes. ITM integrates this perspective by examining how large-scale mental phenomena emerge from the intricate interplay of neurons and synapses. This approach is informed by methods from statistical physics and systems biology, which are used to model the collective dynamics of neural networks.

One key concept in this context is the idea of emergent properties, where the whole is greater than the sum of its parts. In a neural network, the interactions between individual neurons can give rise to emergent phenomena such as consciousness, memory, and perception. These emergent properties are not directly predictable from the properties of individual neurons but arise from the complexity of their interactions.

Mathematically, the emergence of mental states can be modeled using methods from dynamical systems theory and network analysis. For instance, the global behavior of a neural network can be described by a set of coupled differential equations:

$$\frac{d\mathbf{X}}{dt} = F(\mathbf{X}, \mathbf{W}, \mathbf{I}), \tag{4.20}$$

where $\mathbf{X}$ represents the state vector of the network (e.g., the membrane potentials of all neurons), $\mathbf{W}$ is the matrix of synaptic weights, and $\mathbf{I}$ represents external inputs. The function F describes the interactions between neurons and the influence of external stimuli. The solutions to these equations can reveal the emergence of stable patterns of activity, such as synchronized oscillations or chaotic dynamics, which correspond to different mental states.

In the context of AHI development, understanding and leveraging these emergent properties is critical. As artificial components are introduced and integrated into the brain, the overall system's capacity to generate complex mental states must be maintained or enhanced, ensuring that the emergent consciousness remains coherent and continuous.

Identity Element: Linking Mental States to Neural States

Identity theory provides the final element of ITM by asserting that mental states are identical to specific physical states in the brain. This perspective is crucial for understanding the NCC and for bridging the gap between the physical processes in the brain and the subjective experiences that constitute consciousness.

The relationship between a mental state M and its corresponding neural state N can be described as:

$$M \equiv N, \tag{4.21}$$

where $\equiv$ denotes the identity relation. This equation encapsulates the idea that every mental state is identical to a specific neural configuration, which can be measured and analyzed using neuroimaging techniques such as fMRI or electrophysiology.

For AHIs, this identity element is crucial in ensuring that as biological components are replaced with artificial ones, the specific neural states that correspond to an individual's mental states are preserved. This preservation is vital for maintaining continuity of consciousness and personal identity throughout the augmentation process.

Comparing ITM with Information Integration Theory

While ITM offers a holistic and integrative framework for understanding consciousness, it is useful to compare it with other prominent theories, such as the IIT proposed by Giulio Tononi. IIT posits that consciousness arises from a system's ability to integrate information, and it introduces the concept of "phi" (φ), a measure of the amount of integrated information generated by the system. According to IIT, consciousness is an intrinsic property of systems with high (φ), and the level of consciousness is determined by the amount of integrated information. In contrast, ITM does not focus solely on information integration but instead emphasizes the complexity and interconnectedness of different mental states and processes within the brain.

One key difference is that while IIT views consciousness as a fundamental property that can be quantified, ITM sees consciousness as an emergent property arising from the dynamic interactions of neural networks, grounded in both the functional and physical attributes of the brain. This distinction highlights ITM's broader scope, as it seeks to encompass multiple aspects of consciousness, from the physical basis of mental states to their functional roles and emergent properties.

The Holistic Approach of ITM: A Path Forward

The ITM is a distinct approach to understanding consciousness, drawing on the strengths of multiple theories while addressing their individual limitations. By synthesizing elements from functionalism, materialism, reductive materialism, and identity theory, ITM offers a more nuanced and complete explanation of how consciousness arises from the brain's physical and functional properties.

One of ITM's key strengths is its ability to accommodate the complexity of the brain's organization and the emergent nature of consciousness. It recognizes that consciousness cannot be fully explained by any single theory or approach but rather requires an integrative perspective that considers the interplay of multiple factors. It offers a promising framework for understanding the mind and consciousness, one that is grounded in both science and philosophy. By integrating insights from diverse theories, ITM intends to provide a pathway toward a more complete and holistic understanding of the mind, capable of guiding future research and advancing our knowledge of consciousness.

Self-Organization and the Brain's Complexity

The brain's ability to self-organize is a fundamental characteristic that underlies its remarkable complexity and adaptability. Unlike traditional computers, which operate based on fixed algorithms and static architectures, the brain is a dynamic system that continuously reconfigures itself in response to sensory inputs and environmental changes. This capacity for self-organization is crucial for learning, memory formation, and cognitive functions, and it sets the brain apart as an unparalleled computational system.

Neural Networks and Synaptic Plasticity

At the cellular level, the brain is composed of billions of neurons that form intricate and highly interconnected networks. These connections, known as synapses, are not static; they are continuously modified through processes such as synaptic plasticity, which allows the brain to adapt to new information and experiences. Synaptic plasticity refers to the strengthening or weakening of synapses based on the activity patterns of the connected neurons. This adaptability is key to the brain's ability to learn from experience and supports a wide range of cognitive functions.

One of the most well-known mechanisms of synaptic plasticity is Hebbian learning, often summarized by the phrase "cells that fire together wire together." According to Hebb's rule, when a presynaptic neuron consistently activates a postsynaptic neuron, the synaptic connection between them is strengthened. Mathematically, this can be expressed as:

$$\Delta w_{ij} = \eta x_i y_j, \tag{4.22}$$

where Δw_{ij} represents the change in the synaptic weight between neurons i and j, η is the learning rate, and x_i and y_j are the activation levels of the presynaptic and postsynaptic neurons, respectively. This simple yet powerful rule captures the essence of how neural circuits can self-organize based on activity patterns, leading to the formation and refinement of memory traces and other learned behaviors.

A specific manifestation of Hebbian plasticity is long-term potentiation (LTP), a process in which synaptic strength is persistently increased following high-frequency stimulation of the synapse. LTP is considered one of the primary mechanisms underlying learning and memory in the brain. The mathematical modeling of LTP often involves equations that describe the changes in synaptic efficacy over time as a function of neural activity:

$$\Delta w(t) = \alpha \cdot \frac{dV}{dt} + \beta \cdot V(t), \tag{4.23}$$

where α and β are parameters that determine the rate of change in synaptic strength, $V(t)$ is the postsynaptic potential at time t, and $\frac{dV}{dt}$ represents the rate of change of the postsynaptic potential. This model helps to quantify how repeated activation of a synapse can lead to long-lasting changes in its strength, contributing to the stabilization of learning and memory processes.

The Role of the Hippocampus in Memory Formation

The hippocampus, a seahorse-shaped structure located in the medial temporal lobe, plays a crucial role in memory formation and spatial navigation. It acts as a central hub for processing sensory inputs and converting them into long-term memories. The hippocampus evaluates the significance of sensory information, integrates it with existing knowledge, and consolidates it into long-term storage in the neocortex [44].

The process of memory consolidation in the hippocampus can be understood through the concept of neural reactivation. During sleep or rest, the hippocampus "replays" patterns of neural activity associated with recent experiences, which strengthens the synaptic connections involved in those memories. This reactivation is thought to contribute to the transfer of memories from the hippocampus to the neocortex, where they are stored for long-term retrieval.

Mathematically, the process of memory consolidation can be modeled using concepts from dynamical systems theory. The hippocampal activity can be represented by a set of coupled differential equations that describe the evolution of neural states over time:

$$\frac{d\mathbf{X}}{dt} = F(\mathbf{X}, \mathbf{W}, \mathbf{I}) \tag{4.24}$$

where **X** represents the state vector of hippocampal neurons, **W** is the matrix of synaptic weights, and **I** represents external inputs. The function F encapsulates the dynamics of neural interactions, and the solutions to these equations describe the temporal patterns of hippocampal activity during memory processing.

Modular Organization and Cognitive Flexibility

The brain's modular organization is another key factor contributing to its complexity and adaptability. Different regions of the brain are specialized for specific functions, such as vision, language, or motor control, but these regions are also highly interconnected, allowing for efficient information processing and integration. This modular architecture supports the brain's ability to handle complex tasks like reasoning, problem-solving, and decision-making.

Each module or cortical area operates as a semi-autonomous unit, processing specific types of information while interacting with other modules to produce coordinated behavior. For example, the visual cortex processes visual information, while the prefrontal cortex is involved in higher-order cognitive functions such as planning and decision-making. The interaction between these regions allows the brain to integrate sensory inputs with contextual information and past experiences, enabling flexible and adaptive responses to the environment. The mathematical modeling of modular brain networks can be approached using graph theory and network analysis. The brain's modularity can be quantified by metrics such as clustering coefficient, modularity index, and small-worldness, which describe the efficiency and organization of neural networks:

$$Q = \frac{1}{2m} \sum_{ij} \left[A_{ij} - \frac{k_i k_j}{2m} \right] \delta\left(c_i, c_j\right), \tag{4.25}$$

where Q is the modularity index, A_{ij} is the adjacency matrix of the network, k_i and k_j are the degrees of nodes i and j, m is the total number of edges in the network, and $\delta(c_i, c_j)$ is the Kronecker delta, which is 1 if nodes i and j belong to the same module and 0 otherwise. A higher Q value indicates a more modular structure, which is associated with efficient information processing and cognitive flexibility.

Competitive Learning and Self-Organization in Artificial Neural Networks

The principles of self-organization and competitive learning observed in biological neural networks have inspired the development of artificial neural networks (ANNs) that can learn and adapt to new data. Competitive learning algorithms, which mimic the competitive interactions between neurons, are particularly relevant in this context. These algorithms use competition among neurons to determine which neuron will represent a particular input pattern, leading to the formation of specialized feature detectors.

In competitive learning, neurons in a network compete to respond to input patterns, with only the "winning" neuron (the one with the highest activation) being allowed to adjust its weights. This process can be described mathematically by updating the synaptic weights of the winning neuron according to:

$$\Delta w_{ij} = \eta \left(x_j - w_{ij} \right), \tag{4.26}$$

where w_{ij} is the synaptic weight between neuron i and input j, x_j is the input signal, and η is the learning rate. Over time, this leads to the development of distinct neural representations for different input patterns, similar to how the brain forms specialized neural circuits for processing different types of information.

Competitive learning is closely related to Hebbian plasticity; both involve the strengthening of connections based on activity patterns. However, competitive learning introduces an additional layer of complexity by incorporating lateral inhibition, where the activation of one neuron suppresses the activity of neighboring neurons. This mechanism helps to sharpen the network's response to specific inputs and prevents redundant representations.

Hebbian Plasticity and Long-Term Potentiation in Learning

Hebbian plasticity, as mentioned earlier, is a fundamental mechanism underlying learning and memory in both biological and ANNs. In biological systems, Hebbian plasticity is observed in phenomena such as LTP, where repeated activation of a synapse leads to a persistent increase in synaptic strength.

In mathematical terms, Hebbian plasticity can be modeled using equations that describe the change in synaptic weight as a function of the correlation between the presynaptic and postsynaptic activity:

$$\Delta w_{ij} = \eta \cdot \left(x_i \cdot y_j \right), \tag{4.27}$$

where Δw_{ij} represents the change in synaptic weight between neurons i and j, η is the learning rate, and x_i and y_j are the activation levels of the presynaptic and postsynaptic neurons, respectively. This equation encapsulates the essence of Hebbian learning, where the strength of a connection increases proportionally to the degree of correlated activity between the connected neurons.

The application of Hebbian plasticity in ANNs has led to the development of unsupervised learning algorithms, where the network learns to recognize patterns in the input data without explicit labels or supervision. This type of learning is particularly useful for tasks such as clustering, pattern recognition, and feature extraction.

The Role of Microtubules in Self-Organization

The notion of the brain as a self-organizing system is central to understanding its remarkable adaptability and cognitive capabilities. Microtubules, self-assembling protein structures found in neurons, play a crucial role in this process. These structures assist in forming and maintaining synaptic connections, which are essential for learning and memory. Synaptic connections are continuously formed, strengthened, or weakened in response to sensory inputs and environmental stimuli, enabling the brain to adapt dynamically to new experiences and challenges.

Microtubules contribute to this self-organization by supporting intracellular transport, maintaining cellular structure, and interacting with other components of the cytoskeleton. Additionally, they play a pivotal role in regulating neural plasticity, the brain's ability to reorganize itself by forming new neural connections throughout life. This plasticity underlies critical processes such as learning, memory consolidation, and recovery from injury.

While their role in self-organization is well established, microtubules have also been proposed to contribute to cognitive processes through quantum mechanisms. The Orch OR model suggests that quantum coherence within microtubules may be linked to higher-order functions, such as consciousness and decision-making [45]. However, this hypothesis remains a subject of ongoing debate and requires further empirical investigation.

In viewing the brain as a self-organizing system, microtubules serve as a foundational component, enabling the dynamic processes that sustain cognition and adaptation. Their interaction with neural networks exemplifies the intricate interplay between structural and functional elements in the brain.

Plasticity and Adaptability in Brain Function

Another critical aspect of brain function is plasticity, or the brain's ability to adapt and change in response to new experiences and demands. This can take many forms, including restructuring neural connections, forming new synapses, and reorganizing entire brain regions. Plasticity plays a crucial role in the brain's development and adaptation, and it allows the brain to continue learning and adapting throughout life [46].

Plasticity is not only crucial for learning and memory but also for the brain's ability to recover from injury. After damage to a particular brain area, other regions can sometimes take over the lost functions, a process known as neuroplasticity. This adaptability underscores the brain's remarkable ability to reconfigure itself in response to changing conditions and is a testament to its self-organizing capabilities.

Self-Organization in Computational Intelligence

The brain's self-organization is a striking feature, enabling its complexity to emerge not just from the sheer number of neurons but also from their dynamic, causal collaborations through ever-changing neural interconnections. The genetic code may hold the key to evolving such a complex system. Viewing the brain as a system with infinite degrees of freedom, only quantum or potentially genetic algorithms could determine the ideal set of neural interconnections for neuron groups, akin to natural selection and evolution [47].

Recent neurobiological advances have shifted our understanding of the brain, moving away from pure intuition and contemplation. Interestingly, our instincts often clash with scientific findings, leading to questions about the brain's resistance to grasping its organization. Intuition suggests a central hub in the brain where all information converges for conscious processing, potentially handling perception, agency, decision-making, planning, and the concept of self. However, the brain's actual structure, particularly the cerebral cortex, comprises numerous distinct regions performing different functions using similar computational algorithms. This modular design enables information exchange across cortical areas [48].

The brain's modular design facilitates generalization, deliberation, and knowledge encoding and supports consciousness unity. However, connections between brain regions don't strongly endorse the idea of strictly hierarchical processing architectures. Instead, connectivity patterns feature parallelism, communication, and distributedness [48]. For instance, neurons in the visual cortex can directly communicate with neurons in the limbic system or executive regions, and many of these connections are reciprocal. The highly structured and organized connection network exhibits properties of small-world networks, allowing efficient information processing and knowledge storage [49].

Competitive Learning in Artificial Neural Networks

Competitive learning algorithms use competition among neurons to determine which neuron will represent a particular input pattern, leading to the formation of specialized feature detectors. In competitive learning, neurons in a network compete to respond to input patterns, with only the "winning" neuron (the one with the highest activation) being allowed to adjust its weights. This process can be described mathematically by updating the synaptic weights of the winning neuron according to:

$$\Delta w_{ij} = \eta\left(x_j - w_{ij}\right), \tag{4.28}$$

where w_{ij} is the synaptic weight between neuron i and input j, x_j is the input signal, and η is the learning rate. Over time, this leads to the development of distinct neural representations for different input patterns, similar to how the brain forms specialized neural circuits for processing different types of information.

Competitive learning is closely related to Hebbian plasticity, as both involve the strengthening of connections based on activity patterns. However, competitive learning introduces an additional layer of complexity by incorporating lateral inhibition, where the activation of one neuron suppresses the activity of neighboring neurons. This mechanism helps to sharpen the network's response to specific inputs and prevents redundant representations.

Competitive learning in ANNs consists of two main components: a feed-forward excitatory network and a lateral inhibitory network. The excitatory network implements an excitatory Hebb's learning rule, while the inhibitory network implements an inhibitory Hebb's learning rule.

The competitive learning algorithm begins by initializing the weights of all neurons in the network randomly. Then, for each input vector, the algorithm computes the output of each neuron in the network and selects the winning neuron based on the highest output. The weights of the winning neuron are then adjusted to match the input vector, while the weights of the other neurons are adjusted to move them farther away from the input vector. This process is repeated for each input vector in the training set until the network has learned to classify the input vectors correctly.

Synchronization and Dynamic Binding in Neural Networks

Binding distributed effects can be achieved using dedicated anatomical circuits, like connecting outputs from units X and Y to a third unit Z and setting appropriate thresholds for unit Z to activate only when X and Y are active simultaneously. This approach encodes relations in fixed anatomical architectures and communicates through conjunction-specific neurons' responses [50]. However, due to its inflexibility and rigidity, this method can only encode frequently occurring stereotyped relations.

An alternative is expressing relations through dynamic patterns, allowing the representation to remain distributed while still functioning as a cohesive whole, a method called pattern assembly coding. With 10^{11} neurons, each serving as an image and a flexible recombination mechanism, a virtually limitless number of distinct distributed representations can be formed. Representations of novel stimuli, constantly changing real-world conditions, and adaptable motor responses are best implemented through dynamically configured assemblies [51]. In the realm of pattern assembly coding, a conundrum arises: How can we determine which activated neurons in a group genuinely contribute to a specific representation? As assemblies inevitably overlap, the brain needs a clear marker to identify which neurons are truly linked within an assembly. Thus, neurons supporting assembly codes must simultaneously convey two messages. First, they signal the presence of the characteristic they represent by increasing their spike frequency or becoming more active. The hallmark of relatedness among cells in an assembly is the precise synchronization of their spikes, often governed by oscillatory activity [51]. This millisecond-level precision allows for the rapid reconfiguration of assemblies, defining relations with the necessary temporal resolution.

Plasticity and Adaptability in Brain Function

Plasticity is not only crucial for learning and memory but also for the brain's ability to recover from injury. After damage to a particular brain area, other regions can sometimes take over the lost functions, a process known as neuroplasticity. This adaptability underscores the brain's remarkable ability to reconfigure itself in response to changing conditions and is a testament to its self-organizing capabilities.

Self-Organization and Learning

At the cellular level, the cerebrum and nervous system consist of an extensive network of cells known as neurons. While these neurons come in a variety of shapes and types, they operate in a similar fashion. In a self-organized brain, neurons can be considered the individual components that interact to create a global pattern. These neurons have long and short extensions, called axons and dendrites, respectively, which serve to connect them. Dendrites carry electrical potentials toward the cell, while axons transmit them away from it. The dendrite of one neuron connects to the axon of another, separated by a small gap known as the synapse or synaptic gap [52].

To pass information to another neuron, a neuron sends an electrical signal down its axon, which prompts the release of neurotransmitters that cross the synapse to the other cell. This type of interaction is local in nature. However, the connections between neurons are not fixed; they're continuously being strengthened and weakened. This interactive learning serves as the foundation for the learning process.

As the fundamental building blocks of the brain, neurons are anything but simple. The integration of sensory information during learning is all the more extraordinary when considering that two neurons in, for instance, the visual and auditory cortices of the brain are separated by billions of intervening neurons and lack direct synapses to facilitate information integration [52].

Another crucial aspect of learning is the ability to differentiate relevant information from irrelevant data. Hebbian learning plays a role in this process, as does competitive learning. Adaptive resonance theory ANNs employ competitive learning and other methods to overcome the stability-plasticity dilemma and warrant exploration [53]. Many successful models utilize Hebb's learning rule, in which neurons rely solely on local information. These models embody many of the characteristics necessary in self-organized systems. Consequently, most models of the brain hinge on the fundamental assumption of Hebbian synaptic plasticity or Hebbian learning. The brain's self-organization, driven by neural networks, synaptic plasticity, modular organization, and competitive learning, underpins its complexity and adaptability. By understanding these principles, we gain insights into both biological and artificial intelligence, paving the way for more sophisticated AI systems that emulate the brain's remarkable capabilities.

Artificial Neural Networks and the Mind

ANNs are computational models that emulate certain aspects of the brain's structure and function. ANNs are designed as systems of interconnected neurons, much like the human brain, to process and transmit information across layers of connections. However, while ANNs can replicate some functions of the brain, they remain far simpler in complexity and adaptability when compared to the human mind.

Hierarchical Organization: Parallels Between ANNs and the Brain

A key similarity between ANNs and the human brain lies in their hierarchical organization. In both systems, information is processed through multiple layers, with each successive layer performing more abstract computations on the data. In an ANN, the input layer first captures raw data, such as pixels in an image, and transmits this data to subsequent layers. Each hidden layer in the network extracts features of increasing abstraction—edges in the first layer, shapes in the next, and, finally, complex structures such as objects or faces in the deeper layers. This hierarchical processing mirrors the way sensory information is handled in the brain. For example, in the visual cortex, early processing stages (e.g., V1) are responsible for detecting basic visual features such as lines and edges, while later stages (e.g., V4, IT) integrate these features to recognize complex objects.

Mathematically, the operation of a neuron in an ANN can be represented as a weighted sum of its inputs followed by a nonlinear activation function:

$$y_j = f\left(\sum_{i=1}^{n} w_{ij}x_i + b_j\right), \tag{4.29}$$

where y_j is the output of neuron j, x_i represents the inputs to the neuron, w_{ij} are the synaptic weights, b_j is the bias term, and $f(\cdot)$ is the activation function that introduces nonlinearity into the model. This function allows ANNs to learn complex, nonlinear mappings from inputs to outputs, analogous to the way neural circuits in the brain process information.

The success of ANNs in tasks such as image recognition, natural language processing, and speech recognition can be attributed to their ability to exploit this hierarchical structure. However, despite these achievements, ANNs are still vastly inferior to the human brain in several critical aspects.

Complexity and Connectivity: The Scale of the Human Brain

The human brain is an extraordinarily complex organ, consisting of approximately 86 billion neurons, each connected to thousands of others via synapses. This dense network enables the brain to process vast amounts of information in parallel, facilitating highly sophisticated cognitive functions such as reasoning, decision-making, and emotional responses. The connectivity of the brain is often described in terms of a small-world network, where neurons are highly clustered with short path lengths between any two neurons, optimizing both local processing and global communication.

In contrast, even the most advanced ANNs today contain far fewer neurons and synaptic connections. The largest ANNs, such as those used in deep learning models, typically have on the order of millions or billions of parameters (weights), a number that pales in comparison to the trillions of synapses in the human brain. This limitation in scale fundamentally restricts the capabilities of ANNs, particularly in terms of their ability to generalize knowledge across different domains and perform complex, multistep reasoning.

One way to model the network connectivity in the brain is through the use of graph theory, where neurons are represented as nodes and synapses as edges. The efficiency of information processing in the brain can be quantified by measures such as the clustering coefficient C and the characteristic path length L:

$$C = \frac{1}{N}\sum_{i=1}^{N}\frac{2E_i}{k_i\left(k_i - 1\right)}, \tag{4.30}$$

$$L = \frac{1}{N\left(N - 1\right)}\sum_{i \neq j} d_{ij}, \tag{4.31}$$

where E_i is the number of edges between the neighbors of node i, k_i is the degree of node i, and d_{ij} is the shortest path length between nodes i and j.

The small-world nature of the brain's network, characterized by high clustering and short path lengths, allows for efficient and robust information processing—a feature that ANNs, with their limited connectivity, are yet to fully replicate.

Neuroplasticity versus Static Training: Adaptability in the Brain and ANNs

Another significant difference between the brain and ANNs is the brain's remarkable adaptability, known as neuroplasticity. The human brain is capable of continuous learning and reorganization throughout life, adapting to new experiences, learning new skills, and recovering from injuries. This adaptability is driven by mechanisms such as synaptic plasticity, where the strength of synaptic connections changes in response to experience, and structural plasticity, where new synapses form and old ones are pruned based on the brain's activity patterns.

In contrast, most ANNs are trained on a fixed dataset and often struggle to adapt to new data after training is complete. This limitation is particularly evident in the phenomenon of catastrophic forgetting, where an ANN, when trained sequentially on multiple tasks, tends to forget previously learned information upon learning new tasks. This is a stark contrast to the human brain, which can retain and integrate old and new knowledge seamlessly.

Efforts to address catastrophic forgetting in ANNs include techniques such as elastic weight consolidation (EWC), which adjusts the network's parameters to balance the retention of previous knowledge with the acquisition of new information:

$$L(\theta) = L_{\text{new}}(\theta) + \frac{\lambda}{2}\sum_i F_i\left(\theta_i - \theta_i^*\right)^2, \tag{4.32}$$

where $L_{\text{new}}(\theta)$ is the loss function for the new task, θ_i^* represents the optimal parameters for previous tasks, F_i is the Fisher information matrix, and λ is a hyperparameter that controls the trade-off between new learning and memory retention. While such techniques offer some improvements, they do not fully replicate the brain's flexibility and resilience.

Consciousness and Qualia: The Human Mind's Subjective Experience

Perhaps the most profound difference between ANNs and the human brain is the concept of consciousness. These experiences are deeply tied to our perception of the world and our sense of self. The nature of consciousness and whether it can be replicated in machines remains one of the most challenging and enigmatic questions in both neuroscience and AI.

Currently, there is no evidence to suggest that ANNs possess any form of consciousness. While they can process inputs and generate outputs based on learned patterns, there is no indication that they "experience" these processes in the way humans do. The lack of a theoretical framework for understanding machine consciousness further complicates the development of AI systems that might one day exhibit qualities resembling human consciousness.

One speculative approach to achieving AGI involves whole-brain emulation, which entails creating a detailed digital replica of the human brain by replicating its structure, connections, and functions. The idea is that by emulating the brain's architecture and processes, we might be able to create an artificial mind capable of consciousness. However, this approach faces significant challenges, particularly in understanding and replicating the brain's functions at the quantum physical level, where some theories suggest that consciousness and other cognitive functions might involve quantum processes.

Toward AGI: Integrating Quantum Computing and Brain Machine Interfaces

The path to achieving AGI may involve developing neural networks that more closely approximate the scale and efficiency of the human brain. One promising approach is the integration of quantum computing with ANNs. Quantum computing, which leverages the principles of quantum mechanics, has the potential to solve complex problems that are currently intractable for classical computers. Quantum algorithms can explore vast solution spaces simultaneously, providing a significant computational advantage over classical algorithms.

By combining quantum computing with classical ANNs, we could develop hybrid frameworks that enable advanced problem-solving, learning, and adaptation capabilities beyond the reach of classical computing alone. Such a hybrid system might exploit quantum entanglement and superposition to achieve more efficient learning and decision-making processes, potentially moving closer to the development of AGI.

Another promising avenue is the gradual integration of artificial components with the human brain through brain-machine interfaces (BMIs). BMIs create direct communication pathways between the brain and external devices, allowing for the augmentation of cognitive functions and the seamless integration of human and machine intelligence. Over time, these interfaces could enable the development of hybrid systems that combine the best of both worlds—human intuition and creativity with machine precision and computational power.

While ANNs and the human brain share some structural similarities, the differences in complexity, adaptability, and subjective experience are profound. Bridging the gap between artificial and human cognition will require technological advancements and a deeper understanding of the fundamental principles underlying consciousness and intelligence. The future of AI lies in exploring these frontiers, pushing the boundaries of what machines can achieve, and ultimately creating systems that can think and experience the world in ways that are genuinely akin to human minds.

Toward a Unified Theory of Consciousness

In this chapter, we have ventured into the complex landscape of consciousness, exploring its nature through various theoretical frameworks, including materialism, functionalism, identity theory, and quantum mechanics. Each of these perspectives has contributed valuable insights into how consciousness might emerge from the brain's physical and functional properties. However, none of these theories alone can fully capture the multifaceted nature of consciousness—its subjective experiences, its adaptability, and its profound depth.

As we conclude this chapter, we introduce the ITM, a framework that synthesizes the strengths of these existing theories while addressing their individual limitations. The ITM proposes a unified approach to understanding consciousness by integrating materialist, functionalist, reductive materialist, and identity theory elements with emerging quantum considerations. This holistic view allows us to explore the mind as a dynamic, self-organizing system, deeply rooted in both physical and functional realities, and possibly extending into the quantum realm.

The ITM is founded on several core principles that distinguish it from other theories. One of its key tenets is the integration of existing theories into a more comprehensive framework. ITM does not reject the insights provided by materialism, functionalism, identity theory, or quantum mechanics but rather incorporates their strengths to offer a more unified understanding of consciousness. This approach views consciousness as an emergent property that arises from the complex interactions of neural networks, which are grounded in physical processes, functional roles, and potentially quantum effects. By embracing the contributions of each theory, ITM provides a more nuanced and complete explanation of how consciousness emerges from the brain's intricate architecture.

Another foundational principle of ITM is its emphasis on self-organization and complexity. The brain is not a static machine; it is a dynamic, self-organizing system capable of continuous adaptation and reconfiguration. This capacity for self-organization is central to the brain's ability to learn, form memories, and adapt to new experiences. ITM recognizes that mental states emerge from the dynamic interplay of neurons and synapses, where synaptic plasticity—the strengthening or weakening of synapses based on activity patterns—plays a critical role. This self-organizing capability is crucial for understanding how the brain can generate the rich, diverse mental experiences that characterize human consciousness.

In addition to these principles, ITM also considers the possibility that quantum processes within neurons could contribute to consciousness.

While the idea that quantum mechanics plays a role in consciousness remains speculative, it is an intriguing possibility that pushes the boundaries of our current understanding. The Orch OR model, for example, suggests that quantum coherence in microtubules within neurons might play a role in consciousness. Although this theory is still debated, ITM includes quantum considerations to ensure that all potential mechanisms contributing to consciousness are explored.

Furthermore, ITM is particularly relevant to the development of AHI, a concept that envisions humans evolving toward advanced cognitive capabilities akin to AGI, but based on a biological substrate—the brain. ITM provides the theoretical foundation to explore how gradual augmentation and replacement of biological components with artificial ones can lead to enhanced or transformed consciousness, ensuring continuity of identity and self-awareness in AHIs. As we move toward integrating artificial components with human cognition, ITM offers a roadmap to navigate the complexities of maintaining consciousness and identity during this transition.

The ITM also addresses the limitations of existing theories in explaining the subjective aspects of consciousness—what philosophers refer to as qualia. These are the individual, ineffable experiences that make up our conscious life, such as the redness of a rose or the pain of a headache. While materialism and functionalism provide robust frameworks for understanding the brain's computational processes, they often fall short of explaining why and how these processes give rise to subjective experiences. ITM seeks to bridge this explanatory gap by considering the brain as a system where complexity, self-organization, and possibly quantum mechanics come together to produce not just functional mental states but also the rich, subjective experiences that define human consciousness.

Moreover, ITM's emphasis on self-organization aligns closely with the brain's natural processes of learning and adaptation. The brain's ability to reconfigure its neural networks in response to new information is a testament to its plasticity and flexibility. This adaptability is a critical aspect of consciousness, as it allows the brain to navigate a constantly changing environment, form new memories, and even recover from injury. ITM's recognition of these processes underscores the importance of viewing consciousness as a dynamic, evolving phenomenon rather than a static property.

In the context of AI development, ITM provides a valuable framework for creating systems that are not only intelligent but also capable of adapting and evolving over time. Traditional AI systems are often limited by their static architectures and predefined algorithms, which constrain their ability to learn and adapt to new situations. ITM suggests that by incorporating principles of self-organization and complexity, AI systems could become more flexible, adaptive, and capable of handling the complexities of real-world environments. This approach could lead to the development of AI that more closely mirrors the adaptability and resilience of the human brain.

Finally, ITM's holistic approach extends to its consideration of ethical implications in the development of AI and AHI. As we move closer to creating systems that may possess some form of consciousness or cognitive abilities comparable to humans, it is crucial to consider the ethical and societal impact of these technologies. ITM emphasizes the importance of ensuring that the development of AI and AHI aligns with human values and ethical principles, particularly in preserving the continuity of consciousness and identity in augmented or artificial beings.

The ITM offers a comprehensive framework that synthesizes the strengths of existing theories while addressing their limitations. ITM provides a more complete understanding of the mind by viewing consciousness as an emergent property of a dynamic, self-organizing system grounded in physical processes and potentially quantum mechanics. This holistic approach aims to deepen our understanding of consciousness and guides the ethical and responsible development of advanced AI and AHI systems, paving the way for future research and innovation in these fields.

5 Exploring the Human Brain: The Seat of Consciousness

Understanding the intricacies of the human brain is key to unraveling the mysteries of consciousness—the very essence of our awareness and self. Neuroscience has made significant strides in mapping the connections between brain function and body function, offering insights that not only deepen our understanding of human cognition but also guide the development of advanced computational models.

Two primary types of experiments have been particularly valuable in this endeavor:

- Revealing the connections between specific brain segments and the corresponding parts of the body they control
- Uncovering the relationships between groups of neurons and the types of sensory input they respond to

These experiments have shown that the brain is not merely a random collection of neurons but a highly organized network where spatial arrangements play a crucial role. Neurons that manage similar functions tend to be located close to one another, forming specialized clusters that are critical for tasks like pattern classification and sensory processing.

One of the most compelling examples of this organization is the brain's ability to form self-organizing feature maps (SOFM)—a phenomenon that has profound implications for both neuroscience and artificial intelligence. These maps are topographic representations in which neurons responding to similar sensory inputs are spatially arranged in the brain. For instance, in the visual cortex, neighboring neurons respond to neighboring areas of the visual field, creating a detailed map of visual input. This spatial organization is not static; it can adapt and change in response to sensory experiences or environmental changes, demonstrating the brain's remarkable plasticity.

A striking illustration of this adaptability can be found in experiments with somatosensory maps in monkeys. When a finger is removed, the corresponding neural zones in the brain do not remain inactive; instead, the zones for the remaining fingers expand into the area that was previously sensitive to the removed finger. This reorganization highlights the brain's capacity for self-repair and adaptation, ensuring that sensory processing remains efficient even after physical changes.

The visual cortex, in particular, has been the focus of extensive research, leading to a deep understanding of how sensory maps are formed and maintained. The primary visual cortex, for example, is a topographic map where neighboring neurons are tuned to neighboring areas of the visual field. Neurons in this region are also specialized, with some responding more strongly to specific features like orientation, ocular dominance, or temporal frequency. This specialization is not only crucial for visual perception but also serves as a model for understanding how the brain processes other types of sensory information.

To illustrate this, consider the organization of the monkey striate cortex, as shown in Figure 5.1. This image, obtained through optical recording, reveals the intricate patterns of orientation domains and their association with ocular dominance segment boundaries. Each color represents a different orientation preference, while the white lines mark the boundaries between areas of different ocular dominance. These maps are not perfectly regular; they contain singularities or "pinwheels," where the orientation preference changes continuously and circularly, and linear zones where the preference changes gradually. Such complex

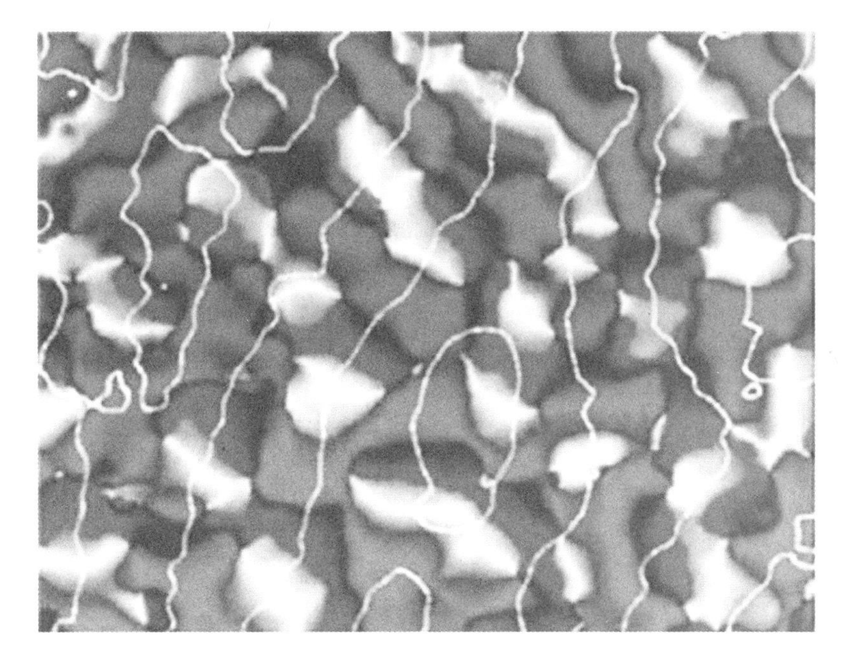

Figure 5.1 Monkey striate cortex recording [54].

patterns are not unique to the visual cortex but are found throughout the brain, reflecting the sophisticated and dynamic nature of neural processing. Crucially, the specialization of neurons in the brain is not fixed or hardwired.

Research has shown that even in adults, the cortex can reorganize itself in response to changes in sensory input or following cortical lesions. This plasticity underscores the brain's ability to adapt to new experiences and recover from injury, a feature that is mirrored in artificial neural networks designed to emulate human cognition.

In computational intelligence, self-organization is a powerful concept that has inspired the development of algorithms like the SOFM. These maps mimic the brain's ability to create spatially organized neural networks, which are particularly useful in tasks like pattern recognition and classification. The development of these algorithms has been influenced by early models of brain function, such as the Von der Malsburg model, which demonstrated how neurons could self-organize based on competitive interactions and Hebbian learning—a process where the connections between neurons strengthen when they are simultaneously active.

The Von der Malsburg model, for instance, showed that neurons could form organized patterns of activity based on local excitatory and long-distance inhibitory connections. This model was groundbreaking in that it introduced the concept of local connectivity—short-distance excitatory connections coupled with long-distance inhibitory connections—that is now a cornerstone of many self-organization models in computational neuroscience.

In this model, the response changes of a neuron over time can be described by the following differential equation:

$$\frac{dH_k(t)}{dt} = -\alpha_k H_k(t) + \sum_{l=1}^{N} p_{lk} H_l^{\star}(t) + \sum_{i=l}^{M} S_{ik} A_i^{\star}(t), \tag{5.1}$$

where $-\alpha_k H_k(t)$ represents the decay of a neuron's response over time, with α_k being the decay constant and $H_k(t)$ the neural response at time t. The term $\sum_{l=1}^{N} p_{lk} H_l^{\star}(t)$ describes the excitation and inhibition the neuron receives from other lateral neurons, where p_{lk} represents the connection strength between cells l and k, and $H_l^{\star}(t)$ results from applying a threshold function to $H_l(t)$. Finally, $\sum_{i=l}^{M} S_{ik} A_i^{\star}(t)$ expresses the effect of retinal cells, with S_{ik} giving the connection strength and $A_i^{\star}(t)$ the stimulus strength.

This model was one of the first to utilize local connectivity and demonstrate the pattern formation that results from short-distance excitatory and long-distance inhibitory connections. It laid the groundwork for modern computational models that seek to replicate the brain's ability to self-organize and adapt to new information.

In this chapter, we will explore how these principles of self-organization and neural plasticity are not only fundamental to understanding the brain but also crucial in the design of artificial systems that aim to replicate or enhance human cognitive abilities. By exploring the mechanisms of neural self-organization, we will uncover how the brain's dynamic and adaptable nature can inspire the next generation of intelligent systems.

Laterally Interconnected Synergetical Self-Organization

Understanding the intricate interplay between neurons is crucial for decoding the brain's complex functions. Neuronal structures feature both lateral and afferent connections, where lateral connections provide inhibitory and excitatory inputs among neighboring neurons, and afferent connections link different layers of neurons. These connections form the backbone of how information is processed and organized in the brain.

The Receptive Field Laterally Interconnected Synergetically Self-Organizing Maps (RF-LISSOM) model builds on this understanding by demonstrating how both lateral and afferent connections can emerge together to represent complex phenomena such as orientation, visual dominance, and selectivity columns. This model is particularly notable because it mirrors the self-organizing capabilities of the human visual cortex, where the brain organizes itself using general learning rules to capture and process sensory information.

LISSOM models strive to replicate the neural structure of the human visual cortex, which self-organizes through general learning rules to capture patterns in both visual inputs and internally generated activities. These models have successfully demonstrated many features observed in human and animal cortexes. At the heart of this approach is the SOFM algorithm, widely known for its applications in data visualization. RF-LISSOM enhances this algorithm by incorporating Hebb's learning rule and parallel connections between neurons, offering a more sophisticated understanding of neural dynamics.

There are several extensions of the original model, such as RF-LISSOM, which focuses on the primary visual cortex (V1). CRF-LISSOM expands this to include the retina and lateral geniculate nucleus, enabling the model to process regular visual stimuli. HLISSOM takes this further by incorporating cortical regions beyond V1, making it capable of modeling not just V1 but the entire mammalian visual system, including complex perceptions like facial recognition.

These models simulate layers of interconnected neurons, where each neuron in an $N \times N$ network receives input from a receptive field of size $s \times s$. The weights of both afferent and lateral connections are initially randomized and are then adjusted through Hebb's learning rules. The model's initial neural response is defined by:

$$n_{ij} = \sigma\left(\sum_{r_1,r_2} \xi_{r_1,r_2} \mu_{ij,r_1,r_2}\right) \tag{5.2}$$

where n_{ij} is the response of neuron (i, j), ξ_{r_1,r_2} represents the activation level of the retinal receptor (r_1, r_2), μ_{ij,r_1,r_2} denotes the strength of the afferent connection, and σ is a function that approximates the sigmoidal activation function.

At each iteration, the neural response is modified to account for lateral interactions:

$$n_{ij} = \sigma\left(\sum_{r_1,r_2} \xi_{r_1,r_2}\mu_{ij,r_1,r_2} + \gamma_e \sum_{k,l} E_{ij,kl}\eta_{kl}(t - \delta t) - \gamma_i \sum_{k,l} I_{ij,kl}\eta_{kl}(t - \delta t)\right) \tag{5.3}$$

The terms represent the contributions of afferent connections, excitatory lateral connections, and inhibitory lateral connections, respectively. The scaling variables γ_e and γ_i control the extent of excitatory and inhibitory influences. Over time, these responses converge to a stable pattern across the topographic map, reflecting the neural organization observed in biological systems.

Subsequent adjustments to both afferent and lateral weights are guided by:

$$w_{ij,mn}(t+1) = \frac{w_{ij,mn}(t) + \alpha\eta_{ij}X_{mn}}{\sum_{mn}\left[w_{ij,mn}(t) + \alpha\eta_{ij}X_{mn}\right]} \tag{5.4}$$

Here, the model employs Hebb's learning rule with normalization, which strengthens connections between neurons with correlated activity while pruning weaker, long-range connections. This pruning is a natural consequence of the model's dynamics, ultimately focusing interactions on nearby neurons as the system trains through numerous iterations.

Despite the complexity of these processes, the RF-LISSOM model effectively captures a wide array of neural patterns and relationships observed in the brain. The study of self-organization in neuroscience continues to be a promising avenue, linking biological understanding with computational modeling in ways that could inform the future of artificial intelligence and neural repair technologies.

Pruning in the Neocortex

The human brain is an astonishingly intricate structure, containing approximately 8.6×10^{10} neurons, each forming about 7,000 synaptic connections with other neurons. As we age, the number of these synaptic connections changes significantly, with a three-year-old child's brain boasting around 10^{15} synapses, while an adult brain contains between 10^{14} and 5×10^{14} synapses [55]. This vast network is central to the brain's ability to learn, adapt, and function. One crucial process in maintaining and optimizing this network is known as pruning—a mechanism that refines neural connections to enhance cognitive efficiency and capacity.

Given the staggering complexity of the neocortex, it's surprising to learn that the human genome, with its 2.9×10^{9} base pairs, encodes a maximum of only 725 megabytes of data. After compression, considering less than 1% variation in individual genomes, this figure reduces to roughly 4 megabytes [56]. This minimal amount of genetic information cannot possibly encode the full complexity of the brain's synaptic networks. Thus, the neocortex likely begins as a collection of numerous smaller, fully connected structures, which are then sculpted and refined through learning processes. Pruning is akin to a sculptor chipping away at a block of marble to reveal the masterpiece within.

In artificial neural networks (ANNs), pruning serves a similar purpose: It removes neurons and connections that do not significantly contribute to the network's overall performance, thereby improving both efficiency and accuracy. In the neocortex, pruning occurs as part of the natural developmental process, streamlining neural circuits for optimal function. Two primary approaches to pruning in ANNs are selective pruning and incremental pruning.

Selective pruning starts with a densely connected network and systematically removes connections and neurons that minimally impact the network's performance. This process is akin to trimming a tree, where only the branches that don't contribute to the tree's growth are removed. As long as the network's error rate remains within acceptable limits, pruning continues, leading to a more efficient structure.

In contrast, incremental pruning begins with a minimal network and gradually adds connections and neurons until the network meets the desired performance threshold. This approach can be more computationally intensive, as it requires constant retraining of the network to evaluate each new structure's effectiveness. However, it can result in a network that is finely tuned to its task. The complexity of the human neocortex is far from being fully explained by our limited genetic code. Pruning processes in the brain and ANNs both serve to refine these networks, making them more efficient and effective. The basis for incremental pruning involves starting with an ANN that has not undergone supervised training, applying a training algorithm for a set number of iterations, and comparing the results against expected error metrics such as the root mean squared error (RMSE). The challenge here lies in the necessity of retraining after each modification, which can significantly increase processing time, especially in complex ANNs designed to model higher cognitive functions of the neocortex.

On the other hand, selective pruning begins with a trained ANN and involves removing connections and neurons individually, assessing whether the resulting error remains within acceptable parameters. If the error is acceptable, the pruning continues. While selective pruning is less time-consuming than incremental pruning, it may not achieve the same level of precision.

Addressing the high computational demands of incremental pruning, particularly in the context of the neocortex's hierarchical organization, requires innovative solutions. One such solution involves leveraging quantum linear superpositions of multiple ANN structures, allowing for parallel exploration of possible configurations. This method considers both spatial and temporal resolution, which is critical when processing real-world inputs in real time—a necessity that goes beyond traditional data mining tasks, such as weather prediction, where training time is less critical.

Ultimately, pruning in both the human brain and ANNs is a vital process for enhancing efficiency, optimizing performance, and ensuring that neural networks, whether biological or artificial, function at their best.

Pruning in Artificial Neural Networks

The complexity of the human brain, with its intricate web of connections and self-organizing capabilities, has long inspired the development of ANNs. Yet, despite these efforts, no single organizing principle can entirely encapsulate the essence of the brain's operation. Instead, a multitude of tools and approaches are required to approximate the brain's functionality, one of which is the concept of self-organization—a process that is crucial not only in biology but also in the advancement of AI systems.

AI systems designed for self-organization aim to mirror the adaptive, dynamic processes observed in the brain. However, a fundamental difference exists between biology and AI, as the requirements and constraints in each field diverge significantly. While feedback mechanisms are integral at various levels within ANNs, not all of these mechanisms lead to true self-organization. For instance, in competitive learning—a key aspect of certain neural networks—feedback occurs during neuron connection and weight adjustment phases. Here, lateral connections between neurons enable the emergence of a "winning" neuron through localized interactions, a hallmark of self-organized behavior. Meanwhile, weight adjustments refine the network by bringing weight vectors closer to the input, thus achieving a form of global learning.

A practical example of this can be found in Adaptive Resonance Theory (ART-1) networks, where interaction occurs between the F1 and F2 modules, both of which are competitive networks. A positive feedback loop operates between the attentional and orientational subsystems within these networks, culminating in a state of resonance that facilitates learning. However, this learning is not considered self-organized in the truest sense, as the resonance state is not inherently generated by any self-organizing component within the network's module structure.

In contrast, SOFMs provide a more direct application of self-organization principles. SOFMs utilize competitive networks to achieve topological ordering, a process where the network organizes itself spatially in response to input. This self-organization occurs at the neuronal level, where a local function governs weight alterations, leading to the emergence of ordered, clustered representations of input data. These clusters are spatially structured, with similar inputs mapped to neighboring nodes, enhancing the network's precision and learning capacity as it processes more data.

One of the strengths of SOFMs, compared to back-propagation networks, lies in their ability to avoid overfitting—a common problem where a network becomes too specialized in the training data, leading to poor generalization on new data. In SOFMs, the ongoing adjustment of neighborhood weights helps maintain a balanced learning process, where the network continues to refine its internal structure without sacrificing accuracy. This process is akin to how the brain continuously adapts and reorganizes itself in response to new experiences, maintaining its ability to learn and function optimally over time.

In essence, while ANNs have made significant strides in mimicking the brain's self-organizing behavior, the gap between AI and biology remains. The exploration of self-organization in ANNs is not only a technical challenge but also a philosophical one, as it raises questions about the nature of learning, adaptation, and the very essence of intelligence—whether biological or artificial.

Beyond Self-Organization

When contemplating the organization of the brain, scientists and philosophers alike have long proposed four different models: the leader, the blueprint, the recipe, and the template. Each model offers a distinct way of understanding how the brain might coordinate its billions of neurons, but each also has its limitations when confronted with the incredible complexity of the human mind.

The first model, the *leader*, imagines that a single neuron or a specific area of the brain acts as a conductor, orchestrating the activity of the rest of the neurons. At first glance, this seems plausible—after all, many systems in nature operate under a central control mechanism. But neuroscience reveals a different story. Imagine a highly specialized neuron, responsible for a crucial function, suddenly being damaged. The result could be catastrophic, akin to losing the conductor of an orchestra mid-performance. However, the brain's resilience lies in its distributed nature; no single neuron or region holds absolute power. Instead, the brain is more like a jazz ensemble, where each musician (or neuron) can take the lead depending on the moment's needs. This multilevel system of organization is evident in how different brain regions become active during specific tasks. For example, the hippocampus, often compared to a librarian, plays a key role in the formation of memories, while the visual cortex specializes in processing visual information. This specialization suggests that self-organization within the brain might not be about a single leader but about regions finding their niche, adapting to the demands placed upon them.

The second model, the *blueprint*, likens the brain's organization to a set of architectural plans encoded in our DNA. This concept is appealing—DNA as the blueprint makes intuitive sense. Yet from a data-theoretical perspective, this model falls short. Consider this: The human genome consists of approximately three billion base pairs, which, while immense, isn't nearly enough to directly encode the trillions of synaptic connections in the brain. Furthermore, even identical twins, who share the same genetic blueprint, develop unique neural architectures. This uniqueness suggests that while DNA sets the stage, it doesn't script the entire play. The brain's connections and capabilities emerge through interactions with the environment, experiences, and learning processes that go far beyond what DNA alone could dictate [57].

The third model, the *recipe*, suggests that the brain could follow a set of instructions, much like baking a cake. But just as with the blueprint model, the brain's complexity outstrips any straightforward set of instructions. While a recipe might tell you the ingredients and steps needed to bake a cake, it doesn't account for variations—such as a different oven or altitude—that might affect the outcome. Similarly, the brain must adapt to countless variables throughout life, making a rigid recipe insufficient to explain its intricate organization.

Finally, there's the *template* model, which proposes that the brain develops according to a predetermined pattern, influenced by nature. Evidence of this can be seen in the brain's somatosensory maps, which reorganize themselves based on sensory input throughout life. These maps, such as those found in the visual cortex, adapt as we learn and grow, responding to the specific visual stimuli we encounter. This plasticity shows that while there might be an initial template guiding brain development, the final structure is shaped by experience. Imagine an artist starting with a rough sketch (the template) but allowing the painting to evolve based on inspiration and the materials at hand. Similarly, the brain's development isn't a fixed process but an ongoing, dynamic interaction with the environment.

Interestingly, this idea of a template influenced by external factors can also be observed in nature, such as in ant colonies. The structure of an ant nest is not encoded in the ants themselves but emerges from their interactions with the environment and each other. Just as ants collectively create complex structures without a central blueprint, the brain's organization emerges through a combination of innate tendencies and environmental influences [58].

In sum, these four models—leader, blueprint, recipe, and template—offer valuable insights but ultimately fall short of capturing the full scope of the brain's complexity. The brain's true organization likely incorporates elements of all these models intertwined in a way that defies simple categorization. The ongoing challenge for neuroscience is to continue unraveling this mystery, exploring how the brain's structure and function emerge from both its biological underpinnings and the rich tapestry of experiences that shape each individual mind.

Interdisciplinary Approaches to Synthetic Brain Augmentation

Synthetic brain augmentation represents a frontier in enhancing human cognition by integrating artificial components with the biological brain. Achieving this ambitious goal requires not only advancements in neuroscience and artificial intelligence but also the convergence of insights and technologies from diverse fields such as genomics, nanotechnology, and biotechnology. This section explores how these disciplines intersect and contribute to the development of synthetic brain augmentation, highlighting the synergies that accelerate progress and the potential for groundbreaking innovations.

The Role of Genomics in Cognitive Enhancement

Genomics, the study of an organism's complete set of DNA, provides the foundational knowledge necessary for understanding the complex genetic networks that regulate neural development, function, and plasticity. Imagine the brain as an intricate orchestra, where each gene acts as a musician playing its part in the symphony of cognition. By mapping the human genome and identifying specific genes associated with cognitive functions, researchers can uncover the genetic basis for critical mental processes such as memory, learning, and neural resilience. This understanding is critical for designing synthetic augmentation strategies that aim to replicate or enhance these natural processes, much like fine-tuning an orchestra to play more harmoniously.

One of the most significant breakthroughs in genomics has been the advent of gene editing technologies, particularly CRISPR-Cas9. This revolutionary tool, often compared to a pair of molecular scissors, allows scientists to make precise alterations to DNA sequences. Imagine being able to edit the genetic code with the same ease as editing text on a computer—deleting, inserting, or correcting letters. CRISPR-Cas9 opens up new avenues for enhancing cognitive functions by targeting specific genes involved in brain activity.

For instance, researchers are exploring the possibility of using CRISPR-Cas9 to upregulate the expression of genes that promote neurogenesis, the process by which new neurons are formed in the brain. Imagine being able to enhance your brain's ability to generate new neurons, much like upgrading the hardware of a computer. This could potentially lead to improvements in memory and learning capabilities, making it easier to acquire new skills or retain complex information. The implications of such enhancements are vast, from students mastering new subjects more quickly to professionals excelling in increasingly demanding cognitive environments.

However, the static genome tells only part of the story. The field of epigenetics—studying how gene expression is regulated without altering the underlying DNA sequence—adds a dynamic layer to our understanding of cognition. Epigenetic modifications, such as DNA methylation and histone modification, can be influenced by environmental factors and life experiences. Think of epigenetics as a set of switches that can turn genes on or off, dimming or amplifying their effects. These modifications are essential for processes like learning and memory consolidation, allowing the brain to adapt to new information and experiences. Consider the case of identical twins, who, despite sharing the same genetic blueprint, can develop distinct personalities and cognitive abilities based on their unique life experiences. This phenomenon can be partly explained by epigenetics, which can cause different patterns of gene expression in each twin, leading to varying levels of cognitive function. In the context of synthetic brain augmentation, epigenetic mechanisms could be harnessed to create systems that are not only genetically optimized but also capable of dynamic adaptation. Imagine an augmented brain that can adjust its cognitive capabilities in response to changing environments or new challenges, much like how our natural brains learn and evolve over time.

The potential for genomics and epigenetics to enhance cognition also paves the way for personalized augmentation strategies. Personalized medicine, driven by genomic data, is increasingly becoming a reality in healthcare, allowing treatments to be tailored to an individual's genetic profile. This personalized approach can be extended to synthetic brain augmentation as well. By analyzing an individual's genetic makeup, researchers can design augmentation strategies that maximize compatibility and efficacy, much like a tailor crafting a bespoke suit that fits perfectly.

For example, a person with a genetic predisposition to neurodegenerative diseases could receive targeted gene therapy to prevent or delay the onset of such conditions while simultaneously benefiting from cognitive enhancements that align with their unique genetic profile. This personalization ensures that synthetic brain augmentation is not a one-size-fits-all solution but rather a tailored intervention that respects the biological uniqueness of each individual.

The possibilities of genomic advancements in cognitive enhancement are profound. They offer a glimpse into a future where we can not only prevent and cure cognitive decline but can also unlock new levels of human potential.

Nanotechnology and Neural Interface Innovations

Nanotechnology, the science of manipulating matter at the atomic and molecular scale, offers unprecedented precision in the design and implementation of neural interfaces—critical components for integrating synthetic elements with biological brain tissue. Imagine being able to interact with neurons at the level of individual synapses, the tiny junctions through which neurons communicate. Nanoscale devices are engineered with this level of detail, facilitating high-resolution communication between artificial and biological elements in ways that were once the realm of science fiction.

Consider the complexity of the human brain, where trillions of synapses transmit signals at incredible speeds. Ensuring that synthetic components can seamlessly integrate into this intricate system requires innovations that go far beyond traditional technology. Nanotechnology provides the tools to create neural interfaces that not only connect with individual neurons but do so with the precision required to maintain the brain's natural functionality. For instance, nanoelectrodes—tiny conductive devices—can be used to record and stimulate neural activity with such precision that they can differentiate between signals from neighboring synapses.

One of the most remarkable applications of nanotechnology in neural interfaces is the development of materials that enhance the quality of neural signal transmission. Efficient signal transmission is a cornerstone of successful synthetic brain augmentation, as it ensures that the communication between synthetic and biological components is both clear and reliable. Imagine trying to hear a conversation in a noisy room—this is similar to the challenge faced by neural interfaces in distinguishing useful signals from background noise. Nanoengineered electrodes can be designed to improve signal-to-noise ratios, much like noise-canceling headphones, allowing for clearer communication and reducing the likelihood of errors.

An exciting example of this innovation is the use of graphene, a single layer of carbon atoms arranged in a hexagonal lattice, known for its exceptional electrical conductivity and strength. Graphene-based electrodes can transmit neural signals with minimal resistance, ensuring that the signals remain strong and clear as they pass through the neural interface. This level of precision is crucial for applications such as deep brain stimulation, where controlling neural activity with high accuracy can lead to significant therapeutic outcomes, such as alleviating symptoms in patients with Parkinson's disease.

Beyond signal transmission, nanotechnology plays a vital role in developing advanced drug delivery systems that support synthetic brain augmentation. Imagine being able to deliver therapeutic agents directly to specific brain regions with pinpoint accuracy, minimizing side effects and maximizing efficacy. Nanocarriers, which are tiny particles designed to carry drugs, can be engineered to navigate through the brain's complex landscape and release their payload only at targeted sites. This targeted approach is especially important for treating conditions like brain tumors or neurodegenerative diseases, where conventional drug delivery methods often struggle to reach affected areas without damaging healthy tissue.

For example, researchers are developing nanoparticles coated with ligands—molecules that bind to specific receptors on the surface of brain cells. These nanoparticles can cross the blood-brain barrier, a protective shield that typically prevents most drugs from entering the brain, and deliver their therapeutic cargo precisely where it's needed. Such precision could revolutionize treatments for conditions like Alzheimer's disease, offering new hope for slowing or even reversing cognitive decline.

One of the most cutting-edge approaches in nanotechnology is the creation of self-assembling nanostructures, which represent a significant leap forward in developing adaptive and resilient neural interfaces. These structures can dynamically organize themselves in response to environmental cues, mimicking the brain's natural processes of synapse formation and reorganization. Imagine a neural interface that, much like a living organism, can adapt to changes in its environment, repairing itself or enhancing its functionality over time.

For instance, self-assembling peptides—short chains of amino acids—can be designed to form nanofibers that promote the regeneration of neural tissue. These fibers can guide the growth of new neurons and synapses, helping to repair damaged brain regions or create new pathways that enhance cognitive function. The ability of these nanostructures to adapt and self-repair makes them ideal for long-term applications in synthetic brain augmentation, where the interface must remain functional and responsive over an extended period.

The potential of nanotechnology to revolutionize neural interfaces cannot be overstated. By enabling precise interaction with individual neurons, enhancing signal transmission, delivering targeted therapies, and creating adaptive structures, nanotechnology is paving the way for synthetic brain augmentation to achieve levels of integration and functionality that were once unimaginable.

Biotechnology Advancements in Neural Repair and Augmentation

Biotechnology has opened new frontiers in the quest to repair and augment the human brain, with tissue engineering leading the charge. Imagine the possibility of growing functional neural tissue in a laboratory, a concept that not long ago seemed like pure science fiction. Today, advances in biomaterials and scaffold design have made it possible to create three-dimensional neural constructs that mimic the architecture and function of native brain tissue. These constructs can be integrated with the brain, offering a new avenue for repairing damage caused by injury or disease, and even enhancing cognitive capabilities.

Tissue engineering involves the use of specially designed scaffolds—three-dimensional structures that provide a supportive environment for cells to grow and organize into functional tissue. For example, a patient who has suffered a traumatic brain injury might benefit from a tissue-engineered graft seeded with neural stem cells. Over time, this graft could integrate with the patient's brain, promoting the regeneration of damaged neurons and restoring lost functions. This approach holds great promise not only for healing but also for augmenting brain function, potentially leading to enhancements in memory, learning, and other cognitive abilities.

Stem cell therapy represents another significant advancement in neural repair and augmentation, offering the tantalizing possibility of replacing damaged or dysfunctional brain tissue with new, healthy cells. Stem cells, which have the unique ability to differentiate into various types of cells, can be used to generate new neurons and glial cells—essential components of the brain's neural circuits. The potential applications of this technology are vast, ranging from treating neurodegenerative diseases like Parkinson's and Alzheimer's to enhancing cognitive functions in healthy individuals.

Consider the story of a patient with Parkinson's disease, a condition characterized by the progressive loss of dopamine-producing neurons in the brain. Traditional treatments, such as medication, can only manage symptoms; they do not address the underlying cause. Stem cell therapy offers a more fundamental solution by potentially regenerating the lost neurons. In clinical trials, researchers have begun using induced pluripotent stem cells (iPSCs), which are derived from the patient's own cells and reprogrammed to develop into dopamine-producing neurons. These neurons can then be transplanted into the brain, where they integrate with existing neural networks and restore dopamine levels, thereby alleviating symptoms and improving quality of life.

In addition to tissue engineering and stem cell therapies, the development of biohybrid systems represents a cutting-edge approach to brain augmentation. These systems combine biological and synthetic components to create hybrid entities that leverage the strengths of both domains. Biological neurons, with their inherent flexibility and adaptability, are ideal for tasks that require nuanced, context-dependent processing. On the other hand, synthetic components excel at tasks that demand speed, precision, and computational power.

Imagine a neural prosthetic device designed to restore vision in a person who is blind. In such a biohybrid system, biological neurons could be used to process visual information in a way that mimics natural vision, while synthetic components, such as microelectronic chips, could enhance the processing power, enabling the brain to handle more complex visual tasks than ever before. This fusion of biology and technology could result in a level of vision enhancement that goes beyond simply restoring sight, potentially allowing individuals to perceive a broader spectrum of light or even see in low-light conditions far better than the natural human eye.

Protein engineering also plays a crucial role in the development of biohybrid systems by designing and synthesizing proteins with specific properties that enhance synaptic transmission and support the long-term integration of synthetic elements into the brain. Proteins are the workhorses of biological systems, facilitating communication between cells and ensuring the proper functioning of neural networks. By engineering proteins that mimic or enhance the function of natural neurotransmitters, scientists can improve the efficiency of communication between synthetic and biological neurons, creating more seamless and effective neural interfaces.

For example, in a biohybrid memory device, engineered proteins could be used to strengthen synaptic connections, thereby enhancing the storage and retrieval of information. These proteins might also be designed to promote synaptic plasticity—the brain's ability to adapt and reorganize itself in response to new information or experiences. This could result in a system where learning and memory processes are not only preserved but also significantly enhanced, enabling users to retain and recall information with greater ease and precision. The advancements in biotechnology, from tissue engineering to stem cell therapies and biohybrid systems, are paving the way for a new era in neural repair and augmentation. These technologies offer the potential not only to restore lost functions but also to enhance the brain's natural capabilities, pushing the boundaries of what it means to be human.

Synergistic Effects and Future Directions

The convergence of genomics, nanotechnology, and biotechnology is more than just the sum of its parts—it is a powerful synergy that is driving the rapid advancement of synthetic brain augmentation. Each of these disciplines contributes unique tools and insights that, when combined, enable the development of neural augmentation systems far beyond what could be achieved by any single field alone. This interdisciplinary collaboration is akin to a symphony, where genomics, nanotechnology, and biotechnology each play their part in creating a harmonious whole, leading to innovations that push the boundaries of human cognition and health.

Consider the analogy of building a complex machine, where each part must fit perfectly with the others to function correctly. Genomics provides the blueprint, offering insights into the genetic foundations of neural function and revealing the specific genes that could be targeted for enhancement. With its unparalleled precision, nanotechnology acts as the machinery, enabling the creation of nanoscale devices that can interact directly with neurons, ensuring that these genetic enhancements are delivered and integrated with the utmost accuracy. Biotechnology, particularly through advances in tissue engineering and stem cell therapy, serves as the finishing touch, providing the biological material and systems necessary to repair, enhance, and sustain the augmented brain.

This convergence is not just theoretical—it is already happening in research labs and clinical trials around the world. For instance, researchers are using genomic data to tailor nanotechnology-based drug delivery systems that target specific neural pathways, enhancing cognitive functions with unprecedented precision. At the same time, biotechnologists are engineering biohybrid systems that combine these advancements, creating neural interfaces that are both adaptive and resilient, capable of evolving alongside the biological brain.

The development of synthetic brain augmentation technologies is not confined to the ivory towers of academia. It requires a collaborative effort across multiple fields and industries, bringing together the brightest minds from genomics, nanotechnology, biotechnology, neuroscience, and artificial intelligence. Innovation ecosystems—networks of researchers, companies, and institutions—are essential for driving progress. These ecosystems foster a culture of collaboration and knowledge exchange, where ideas can cross-pollinate and lead to breakthroughs that would be impossible within the confines of a single discipline.

Imagine a conference room where a geneticist, a nanotechnologist, and a neuroscientist are brainstorming together. The geneticist explains how a particular gene could enhance synaptic plasticity, while the nanotechnologist suggests a nanoscale delivery system to target this gene precisely within the brain. The neuroscientist then adds insights on how this enhancement could be integrated with existing neural networks to improve memory and learning. This kind of interdisciplinary dialogue is happening in real time, accelerating the pace of discovery and bringing us closer to practical applications of synthetic brain augmentation.

As we look to the future, several key areas of research will be critical in moving from experimental evidence to practical implementation:

- **Integration of Genomic Data:** Genomics offers a treasure trove of data that, when harnessed correctly, can inform the design of personalized augmentation strategies tailored to the genetic profiles of individual users. By understanding the genetic factors that influence cognition, researchers can develop targeted interventions that maximize the efficacy of brain augmentation technologies, ensuring that each person receives a treatment optimized for their unique genetic makeup.

- **Advances in Nanotechnology:** The future of neural interfaces lies in the continued development of sophisticated nanoscale devices that can interact seamlessly with the brain. These advancements will include more refined electrodes with better signal-to-noise ratios, as well as nanocarriers that deliver therapeutic agents directly to specific brain regions. Such technologies will be essential for enhancing the performance and longevity of synthetic brain augmentation systems.
- **Biotechnology Innovations:** Tissue engineering, stem cell therapies, and protein engineering are at the forefront of creating biohybrid systems that merge synthetic and biological components. As these technologies evolve, they will allow for the seamless integration of synthetic elements into the brain, creating augmentation systems that not only replicate but also enhance the brain's natural capabilities.
- **Ethical and Regulatory Frameworks:** With great power comes great responsibility. As we develop these groundbreaking technologies, it is imperative to establish comprehensive ethical guidelines and regulatory frameworks that ensure advancements in brain augmentation are pursued responsibly. This includes respecting individual autonomy, ensuring equitable access, and addressing the societal impact of these technologies.

The future of synthetic brain augmentation is marked by the potential to not only replicate the capabilities of the biological brain but also to transcend them. We are on the brink of transforming human cognition in ways that were once the stuff of science fiction.

Philosophical Perspectives on Synthetic Brain Augmentation

The advent of synthetic brain augmentation technologies challenges some of the most fundamental philosophical questions about identity, consciousness, and what it means to be human. As we push the boundaries of human cognition through technological enhancements, we must consider the deep implications of these advancements. This section explores the philosophical perspectives that arise from synthetic brain augmentation, offering insights into the continuity of identity, the redefinition of consciousness, the concepts of personhood and humanity, and the ethical and societal implications of these technologies.

The Continuity of Identity

The continuity of identity is one of the most pressing philosophical issues raised by synthetic brain augmentation. As individuals gradually replace biological components of their brains with artificial counterparts, profound questions arise about whether their personal identity remains intact. This dilemma touches the core of what it means to be "you"—the self that experiences the world, remembers the past, and envisions the future.

Imagine waking up one day to find that parts of your brain have been replaced with synthetic enhancements. Your thoughts flow more quickly, your memory is sharper, and your cognitive abilities are vastly improved. Yet you might wonder: Am I still the same person? If so, what is it that preserves my identity across this transformation? This question is not merely theoretical; it strikes at the heart of human experience and our understanding of selfhood.

Philosophers have grappled with the problem of personal identity for centuries, offering various theories to explain what constitutes the self. These theories provide different perspectives on how identity might be preserved—or altered—in the face of brain augmentation:

- **Psychological Continuity Theory:** This theory posits that personal identity is maintained through the continuity of psychological states, such as memories, beliefs, and personality traits. From this perspective, as long as the augmented individual retains their psychological continuity, their identity remains intact, even if their brain is partially or wholly artificial. For example, if you can still recall your childhood memories, maintain your beliefs, and recognize your personality traits, then you are still "you," regardless of the physical changes to your brain. This theory suggests that the essence of who we are lies in the ongoing narrative of our psychological experiences.
- **Biological Continuity Theory:** According to this theory, personal identity is grounded in the biological continuity of the organism. The replacement of biological components with artificial ones could

therefore challenge the continuity of identity, raising the possibility that the augmented individual is no longer the same person. For instance, if a significant portion of your brain is replaced with synthetic materials, this theory might argue that the "you" that existed before the augmentation is no longer present. The biological connection to your previous self has been severed, leading to a potential break in identity.

- **Narrative Identity Theory:** This theory suggests that personal identity is constructed through the narrative of one's life, encompassing the individual's experiences, relationships, and personal development. Synthetic augmentation might influence the individual's narrative but does not necessarily disrupt the continuity of identity as long as the person perceives themselves as the same protagonist in their life story. Imagine your life as a storybook, with each chapter representing different phases of your existence. Even if the tools you use to think and feel change, as long as you continue to see yourself as the author of your story, your identity remains consistent. The narrative identity theory emphasizes the subjective experience of being the same person over time, despite external changes.

The implications of these theories for synthetic brain augmentation are profound. They force us to reconsider the nature of the self and the criteria by which identity is preserved or altered. For instance, could someone who undergoes significant augmentation still be held accountable for actions taken before the augmentation? Would their relationships and social roles remain the same, or would they need to be renegotiated?

Coming back to the experiment "Ship of Theseus": If every part of a ship is replaced, piece by piece, is it still the same ship? Similarly, if every neuron in your brain is replaced over time with artificial counterparts, are you still the same person? Or does the continuity of self lie in something beyond the physical components—in the patterns, memories, and experiences that make up your life?

These questions are not just philosophical musings; they have real-world implications as we move closer to a future where synthetic brain augmentation becomes a reality. Understanding the continuity of identity is crucial for developing ethical guidelines, legal frameworks, and social norms that will govern this brave new world. It challenges us to think deeply about what it means to be human and how we define the self in an age of unprecedented technological transformation.

Redefining Consciousness

Consciousness, often regarded as the most enigmatic aspect of human existence, lies at the heart of our self-awareness, our perception of the world, and our sense of reality. Traditionally, consciousness has been understood as a phenomenon that arises from the biological processes of the brain—the result of complex interactions among billions of neurons. However, the advent of synthetic brain augmentation challenges this deeply ingrained view by introducing artificial components into the mix, prompting us to ask: Can consciousness exist within an augmented brain, or even within entirely synthetic systems?

Consider a scenario in which a significant portion of your brain is augmented with advanced synthetic components designed to enhance cognitive functions. You still experience thoughts, emotions, and sensations just as before, but now with greater clarity and speed. This raises a profound question: Is your consciousness—the "you" that observes and interacts with the world—still the same? Or has the introduction of synthetic elements altered the very fabric of your conscious experience?

The concept of artificial consciousness, or the potential for synthetic systems to generate conscious experiences, introduces critical philosophical questions that challenge our understanding of what it means to be conscious. Can artificial components, such as those used in brain augmentation, truly replicate or even enhance consciousness? If so, what criteria must these components meet to be considered conscious?

To explore these questions, we can draw upon several philosophical perspectives, each offering a different lens through which to view the relationship between synthetic augmentation and consciousness:

- **Functionalism:** Functionalism offers a compelling approach to understanding consciousness in the context of synthetic augmentation. This theory posits that mental states, including consciousness, are defined by their functional roles rather than by the physical substance that constitutes them. From a functionalist perspective, if synthetic components can perform the same functions as biological neurons—processing information, generating responses, and supporting cognitive processes—then these components might also replicate consciousness. For example, if a synthetic neuron processes sensory input and contributes to your perception of the world in the same way a biological neuron would, functionalism suggests that this process could be part of a conscious experience.

- **Emergentism:** Emergentism provides another intriguing perspective, suggesting that consciousness arises as an emergent property of complex systems. Just as life emerges from the intricate interplay of biological molecules, consciousness might emerge from the complex interactions within neural networks—whether those networks are biological, synthetic, or a combination of both. If synthetic systems can achieve a similar level of complexity as the biological brain, they could potentially generate conscious experiences. This view invites us to consider the possibility that consciousness is not tied to the specific material of the brain but to the complexity and organization of the system as a whole. Imagine a synthetic brain augmentation system that, through its intricate design, gives rise to a conscious mind—a mind that perceives, thinks, and feels in ways that are remarkably human.
- **Phenomenology:** Phenomenology, with its focus on the subjective experience of consciousness, brings us to the heart of what it means to "feel" conscious. This perspective raises questions about whether artificial systems can truly replicate the qualitative aspects of consciousness, known as "qualia." Qualia refer to the individual, subjective experiences that make up our conscious life—such as the redness of a sunset, the taste of chocolate, or the feeling of joy. Phenomenologists might argue that even if synthetic systems can replicate the functional and emergent aspects of consciousness, they may still lack the ability to generate true qualia. This leads to a deeper inquiry: If it could exist, would synthetic consciousness be indistinguishable from human consciousness? Or would it be a different kind of experience altogether—one that we, as biological beings, might not fully understand or relate to?

These perspectives invite us to rethink the very nature of consciousness and to consider the implications of creating or enhancing consciousness through synthetic means. If consciousness can indeed arise within synthetic systems, we must confront new ethical, philosophical, and societal questions. For instance, what rights and responsibilities would a conscious synthetic being have? How would we relate to such entities, and how would they perceive us?

Moreover, the possibility of synthetic consciousness challenges our understanding of human identity and the boundaries of what it means to be alive. It blurs the line between the organic and the artificial, pushing us to reconsider the essence of our existence in a world where consciousness might no longer be the sole domain of biological beings.

It is crucial to engage in deep philosophical reflection. The exploration of artificial consciousness is not just a theoretical exercise; it is a necessary step in preparing for a future where the boundaries of consciousness are expanded beyond the confines of biology, potentially transforming our understanding of life itself.

Concepts of Personhood and Humanity

Personhood is a concept that touches upon the very core of what it means to be an individual in society. It encompasses the qualities that define a person, including consciousness, self-awareness, autonomy, and moral agency. Traditionally, these attributes have been understood as inherently tied to the human experience—rooted in our biological makeup and shaped by our social and cultural contexts. However, synthetic brain augmentation challenges these traditional notions by introducing entities that may possess some or all of these qualities, yet differ fundamentally from unaugmented humans in their composition and capabilities.

Imagine a future where individuals with augmented brains can process information at speeds unimaginable today, possess memories that never fade, and interact with the world through enhanced sensory experiences. These beings might share many characteristics with unaugmented humans—consciousness, rationality, and the ability to make autonomous decisions—yet their enhanced capacities would set them apart. This scenario raises profound questions: Do these augmented beings qualify as persons in the same way that we do? What rights and responsibilities should they have?

Philosophical debates on personhood have long focused on the criteria that must be met for an entity to be considered a person. These criteria typically include the following:

- **Rationality:** The capacity for logical thought, reasoning, and problem solving
- **Autonomy:** The ability to make independent decisions and take actions based on one's own will
- **Moral Agency:** The capacity to understand moral principles and act in accordance with them
- **Self-Awareness:** The ability to reflect on one's own existence, thoughts, and experiences

As synthetic brain augmentation blurs the line between human and machine, it complicates our understanding of these criteria. Consider an augmented individual whose enhanced cognitive abilities allow them to solve complex ethical dilemmas with unprecedented clarity. Their autonomy might be amplified by their augmented capacity to foresee the consequences of their actions, yet this very augmentation may raise doubts about the authenticity of their moral agency. Are they still exercising free will, or are their decisions influenced by the synthetic components within their brain?

This ambiguity extends to the concept of humanity, which is closely tied to personhood but also encompasses broader cultural, social, and ethical dimensions. The idea of humanity has traditionally been rooted in our shared biological heritage—our flesh and blood, our genetic code, and the evolutionary history that has shaped us. But as we introduce beings that possess humanlike consciousness and capabilities yet are fundamentally different in their biological makeup, we are forced to reconsider what it means to be human.

Two contrasting perspectives help to illustrate the ethical and philosophical tensions that arise in this context:

- **Transhumanism:** The transhumanist perspective embraces the idea that humanity can and should transcend its biological limitations through technology. Transhumanists view synthetic brain augmentation as a natural and desirable step in human evolution, one that will lead to the emergence of "post-human" beings with enhanced cognitive, emotional, and physical capabilities. From this viewpoint, the augmentation of the brain is not just an enhancement of human abilities but a transformative process that could fundamentally change our species. The possibility of creating beings with intelligence far beyond our own, who can live indefinitely and experience the world in ways we cannot yet imagine, is seen as a path to fulfilling our potential as a species. However, this perspective also raises concerns about the loss of traditional human experiences and the potential creation of a new elite, further deepening social divides.

- **Human Exceptionalism:** In contrast, the view of human exceptionalism holds that there is something inherently unique and valuable about being human, rooted in our biological nature, cultural heritage, and moral agency. Proponents of this view may see synthetic augmentation as a threat to the very essence of humanity. They might argue that our value lies not in our cognitive abilities or our capacity for technological enhancement, but in our shared human experiences—our capacity for empathy, our relationships, and our ability to navigate the moral complexities of life. From this perspective, synthetic brain augmentation risks eroding the qualities that make us truly human, potentially leading to a future where the line between person and machine is so blurred that we lose sight of what it means to be human at all. These perspectives highlight the ethical and philosophical tensions that arise as we consider the impact of synthetic brain augmentation on our understanding of humanity. Are we on the brink of a new era of human evolution, where technology will enable us to transcend our biological limitations and achieve unimaginable heights? Or are we risking the loss of our humanity by pursuing enhancements that could fundamentally alter who we are?

Ethical and Societal Implications

The ethical implications of synthetic brain augmentation are as vast as they are complex, presenting a series of moral dilemmas and societal challenges that we are only beginning to grapple with. As technology brings us closer to creating beings with cognitive abilities far beyond those of unaugmented humans, we must carefully consider the responsibilities that accompany such profound advancements.

- **Autonomy and Consent:** One of the foremost ethical concerns in synthetic brain augmentation revolves around the autonomy and consent of individuals undergoing these procedures. In a world where the allure of enhanced cognitive abilities might be overwhelming, it is crucial to ensure that individuals fully understand both the risks and benefits of augmentation. The possibility of coercion—whether through social pressure, economic incentives, or subtle psychological manipulation—poses a significant threat to true autonomy. Imagine a future where employers subtly pressure their employees to undergo cognitive enhancements to remain competitive, or where societal expectations create an implicit demand for augmentation. Such scenarios underscore the importance of safeguarding individual consent, ensuring that the decision to augment one's brain is made freely and with a full understanding of the potential consequences.
- **Justice and Accessibility:** The issue of justice and accessibility is another critical concern. If synthetic brain augmentation technologies remain accessible only to the wealthy or privileged, they could exacerbate existing social and economic disparities. We risk creating a society where cognitive enhancement becomes a marker of class distinction, leading to a divide between the "augmented" and the "non-augmented." The implications of such a divide are profound: Those with enhanced cognitive abilities might dominate high-status jobs, accumulate more wealth, and exert greater influence over societal norms and policies. This could lead to a self-perpetuating cycle of inequality, where the benefits of augmentation are concentrated in the hands of a few, while the rest are left behind. To prevent such an outcome, it is essential to consider policies that ensure equitable access to these technologies, perhaps through public funding, subsidies, or regulations that promote fairness and inclusivity.
- **Impact on Society:** Beyond individual ethics, the broader societal impact of synthetic brain augmentation must also be considered. The widespread adoption of these technologies could lead to shifts in social norms, values, and institutions. For example, as more people

become augmented, traditional measures of intelligence, competence, and worth may be redefined. Educational systems might need to adapt to accommodate enhanced cognitive abilities, while job markets could be transformed as certain skills become obsolete and new ones emerge. Moreover, the very definition of what it means to be human could be challenged. As the boundaries between human and machine blur, we may need to rethink concepts like citizenship, rights, and responsibilities, especially as they pertain to augmented individuals. These shifts could lead to significant changes in social hierarchies and power dynamics, raising important questions about how to manage and regulate a society where cognitive enhancement is commonplace. Given the profound implications of synthetic brain augmentation, establishing robust regulatory frameworks is not just important—it is essential. These frameworks must address a range of issues to ensure the safe and ethical development and deployment of augmentation technologies.

- **Safety and Efficacy:** Ensuring that augmentation technologies are both safe and effective is paramount. Rigorous testing and oversight are needed to prevent harmful side effects and to ensure that the promised benefits of augmentation are realized.
- **Privacy and Data Security:** The integration of synthetic components into the brain raises significant concerns about privacy and data security. Protecting the sensitive neural data of augmented individuals from unauthorized access or misuse is crucial in maintaining trust and safeguarding personal autonomy.
- **Rights and Responsibilities:** Defining the rights and responsibilities of augmented beings will be a central challenge. As these individuals may possess cognitive abilities and experiences that differ significantly from those of unaugmented humans, there will be a need to consider new legal and ethical standards that address their unique status.

The philosophical implications of synthetic brain augmentation are indeed vast and multifaceted. They touch on fundamental questions of identity, consciousness, personhood, and ethics, challenging us to rethink our understanding of what it means to be human in an era of unprecedented technological change.

From Neuroscience to Synthetic Brain Augmentation

With its extraordinary structure and function, the human brain serves as both a model and a muse for the development of synthetic brain augmentation technologies. Insights gleaned from neuroscience, particularly in the study of spatial organization and self-organization, are not just academic curiosities—they are the blueprints for revolutionizing human cognition through synthetic means. As we decode the mysteries of the brain, we open the door to possibilities that once belonged solely to the realm of science fiction.

One of the most compelling discoveries in neuroscience is the existence of tonotopic, somatosensory, and retinotopic maps—these are essentially the brain's internal cartography, where different areas are dedicated to processing specific types of sensory information. Imagine a finely tuned orchestra, where each instrument contributes to a harmonious whole. Similarly, these neural maps are meticulously organized to respond to environmental stimuli, constantly adapting and fine-tuning themselves. This dynamic adaptability is not just fascinating; it's instructive. It suggests a model for synthetic neural networks (SNNs) that can learn, adapt, and evolve in response to their environment. By implementing topographic mapping algorithms inspired by these biological phenomena, we can create SNNs that mimic the brain's inherent plasticity—systems that are not just programmed but are capable of growth and transformation. Further exploration of the brain's feature preferences, particularly in the visual cortex, has revealed how neurons specialize in processing specific aspects of sensory input, such as orientation or motion. This specialization is not arbitrary; it is the result of a complex interplay between genetics, environment, and experience. For those of us designing computational models to replicate or even enhance human cognition, this is a treasure trove of information. It allows us to guide the creation of SNNs that don't just process information but do so in a way that mimics the complex, emergent behavior of the human brain. For example, the interplay between ocularity and orientation preferences in neurons provides a framework for developing networks that are not only efficient but also capable of sophisticated pattern recognition and decision-making.

One of the brain's most remarkable features is its ability to reorganize itself—a phenomenon known as neuroplasticity. Whether in response to sensory input or injury, the brain's capacity to adapt is nothing short of miraculous.

Consider the somatosensory maps, which can reorganize themselves following an injury, ensuring that the brain continues to function effectively despite significant changes. This adaptability is a key principle for synthetic brain augmentation. By designing systems that can prune and strengthen synaptic connections based on usage and experience, we can create synthetic systems that are as dynamic and efficient as their biological counterparts. These systems will not be static machines but living entities that grow, adapt, and improve over time, much like the human brain [59].

The journey from understanding the brain to realizing synthetic brain augmentation is one of the most exciting frontiers in modern science. It is a journey that promises not merely to replicate human cognition but to expand it in ways we are only beginning to imagine. The integration of neuroscience insights into the development of SNNs and augmentation technologies holds the key to unlocking new dimensions of human intelligence, creativity, and experience. Imagine a world where our cognitive abilities are no longer constrained by the biological limitations of our brains—where we can enhance memory, accelerate learning, and even interface directly with machines in ways that feel as natural as thought itself.

By harnessing the principles of neural organization, plasticity, and computational modeling, we are aiming to enhance human cognition in ways that were once the stuff of dreams. This endeavor challenges our understanding of consciousness and identity, pushing us to redefine what it means to be human. But beyond these philosophical challenges, it offers something even more profound: the opportunity for unprecedented growth and exploration. As we stand at the cusp of this new era, the question is not just what we will create but what we will become.

Visionary Perspectives on Synthetic Brain Augmentation

The advent of synthetic brain augmentation technologies heralds not just an era of enhanced cognitive abilities but a fundamental redefinition of what it means to be human. This visionary perspective explores the far-reaching implications of these advancements, offering a glimpse into a future where the limitations of our biological origins are transcended, and human potential reaches unprecedented heights.

Imagine a world where cognitive capabilities are not constrained by the physical brain but are amplified beyond recognition. Synthetic brain augmentation promises to extend memory, accelerate learning, and improve decision-making to levels we can scarcely comprehend today. But this is only the beginning. Beyond enhancing our current abilities, synthetic augmentation opens the door to entirely new forms of intelligence and interaction with the world—intelligences that may operate on planes of thought we have yet to conceive.

Consider, for instance, the realm of creativity. Augmented humans could create art, literature, and music with a depth and complexity that far exceeds anything imaginable with our current cognitive frameworks. Imagine a composer not just writing a symphony but simultaneously perceiving and manipulating every note, every instrument, and every emotion it evokes in real-time, across multiple dimensions of sound and experience. Or an artist who can visualize and render entire new worlds, complete with their own physics, flora, and fauna, in the blink of an eye. Such creative feats would redefine the boundaries of human expression, giving rise to cultural and artistic movements that push the limits of imagination.

One of the most captivating prospects of synthetic augmentation lies in its implications for space exploration. Freed from the constraints of biological bodies, augmented humans could embark on interstellar journeys that span thousands of years, perceiving these voyages as mere moments or experiencing them in a condensed, accelerated fashion. Consider the explorers of the future, capable of charting distant galaxies, interacting with alien life forms, and uncovering the mysteries of the universe—all while maintaining a continuity of consciousness that defies the passage of time [60]. With such capabilities, space travel would no longer be a venture for the few but a collective endeavor, expanding humanity's reach into the cosmos.

The potential of synthetic brain augmentation extends even further when we consider the possibility of existing simultaneously in multiple vessels. Imagine a world where your consciousness can inhabit several bodies—whether biological clones or entirely artificial constructs—allowing you to experience life from multiple perspectives at once. You could explore the depths of the ocean in one body while walking on Mars in another, all the while contributing to a global network of shared knowledge and experience. This multiplicity of existence would foster a profound connection with the universe and with each other, enabling us to experience reality in ways that are currently unimaginable [61]. At the heart of synthetic brain augmentation lies perhaps the most profound promise of all: immortality. No longer bound by the physical decay of the biological body, augmented humans could achieve a continuity of consciousness that spans centuries, even millennia. This form of immortality would not just preserve individual experience but allow for the continuous accumulation of knowledge and wisdom, passed down through generations of augmented beings. Imagine a world where the greatest minds of humanity never fade away but continue to contribute to our collective understanding, guiding us through the challenges and opportunities of an ever-evolving existence. It represents the ultimate liberation from the constraints of time, opening endless possibilities for growth, exploration, and the deepening of our understanding of the cosmos. The possibilities of synthetic brain augmentation call us to envision a world where cognitive enhancement knows no bounds, where new frontiers in space and consciousness are explored, and where immortality becomes a tangible reality.

In this future, the full potential of human creativity, intelligence, and spirit can be unleashed, heralding a new era of discovery and existential fulfillment.

Charting the Future of Synthetic Brain Augmentation

The pursuit of synthetic brain augmentation is grounded in a growing body of experimental evidence that underscores the potential of this technology to transform human capabilities. As we explore the mysteries of the brain and the frontiers of computational and materials science, the path toward realizing synthetic brain augmentation becomes increasingly tangible.

Recent advances in brain-machine interfaces (BMIs), neuroprosthetics, and optogenetics offer compelling glimpses into the future of brain augmentation. BMIs that enable direct communication between the brain and external devices have successfully restored sensory and motor functions in individuals with disabilities, demonstrating the feasibility of integrating artificial systems with biological neural networks. Similarly, neuroprosthetics have begun replicating or augmenting cognitive functions such as memory, showcasing the potential for direct enhancement of brain capabilities.

Optogenetics, a technique that uses light to control neurons genetically modified to express light-sensitive ion channels, has provided invaluable insights into the functioning of neural circuits. By enabling precise manipulation of neural activity, optogenetics paves the way for understanding and replicating the complex dynamics of consciousness and cognition in synthetic systems [62].

As the field advances, several key areas of research will be critical in moving from experimental evidence to practical implementation:

- **Material Science Innovations:** Developing biocompatible materials that can seamlessly integrate with the brain's neural tissue is essential for creating synthetic neurons that can replace or augment biological ones without adverse effects
- **Quantum Computing and AI:** Leveraging the computational power of quantum computing and advances in artificial intelligence to design SNNs capable of emulating the complexity and adaptability of the human brain
- **Ethical and Regulatory Frameworks:** Establishing comprehensive ethical guidelines and regulatory frameworks to ensure that advancements in brain augmentation are pursued responsibly, with respect for individual autonomy and societal impact

The journey toward synthetic brain augmentation is both a scientific endeavor and a philosophical exploration into the future of human evolution. The possibilities for enhancing human intelligence, creativity, and experience seem limitless. The challenges ahead are substantial, but so are the opportunities—for extending human life, exploring the universe, and ultimately, understanding the nature of consciousness.

6 The Dual Nature of Consciousness: Algorithmic Precision or Quantum Mystery?

The principles of quantum physics may play a role in how information is processed in the human brain. This is due to the characteristics and behavior of neurons, which are cells in the brain with branching dendrites and axons that transmit signals to other neurons. When a specific neuron, known as the postsynaptic neuron, receives signals from other neurons, called presynaptic neurons, its electrical potential may increase. At rest, the potential of a neuron is typically around −70 mV, but it can be influenced by the signals it receives from other neurons. Each neuron also has a threshold of around −55 mV, which may be exceeded if multiple presynaptic signals raise the neuron's potential. However, this may not always happen, as the synapse transmitting the signal can also be inhibitory, which decreases the postsynaptic neuron's potential and the likelihood of firing.

There have been two principal approaches to understanding the way the human brain processes information: the neurologist's approach, which involves gathering detailed knowledge about the brain and how it processes information; and the information theorist's approach, which consists of understanding the essential functions of the brain and trying to imitate them using computer science techniques like artificial neural networks.

However, there needs to be more than these approaches, because the neurologist's approach only considers macroscopically visible information processing using Newtonian physics. In contrast, the information theorist's approach may consider quantum physics in a quantum artificial neural network (QNN) (i.e., in how quantum computers can be used to more efficiently process classical structures), but still ignoring the role quantum physics may play for processing signals and storing information in the human brain. A complete understanding of the complex process of information processing in the brain must incorporate all three of these areas: the neurologist's knowledge of the brain's inner workings, the computer scientist's knowledge of digital information processing and quantum information theory, and the physicist's knowledge of quantum mechanics and quantum field theory. This integrated approach, known as neuro-engineering, has proved to be helpful in understanding and enhancing the human brain's capabilities.

It is not unreasonable to consider the possibility of quantum effects in the human brain, even though it may seem unlikely at first. There are already some similarities between quantum mechanics and certain aspects of brain function, such as the brain's ability to process information in parallel, its nondeterministic processes, and the reduction of information to single events that can be perceived as conscious experiences.

Quantum physics, with its concepts of entanglement, superposition, deterministic evolution but nondeterministic state-vector reduction, and interference, among others, seems to be a promising candidate for explaining how the brain processes information. It is not a new idea, either. Niels Bohr, one of the pioneers of quantum mechanics, was skeptical about the idea that biological processes might involve quantum physical effects; however, his skepticism would paradoxically influence the founders of quantum biology [63]. In 1967, a hypothesis about how quantum field theory might apply to the human brain was proposed [64]. It suggests that spontaneous symmetry violations could lead to quantum states in the brain called Goldstone bosons, theoretical particles that are associated with the spontaneous breaking of continuous symmetries. They play a crucial role in the Higgs mechanism, which is responsible for endowing fundamental particles with mass. The Higgs boson is a type of Goldstone boson. In addition to the Higgs boson, other types of Goldstone bosons are predicted to exist, and their discovery could help shed light on some of the most profound mysteries of the universe, such as the nature of dark matter and the unification of the fundamental forces. They are long-range correlated waves, and symmetry violation has been interpreted as a natural measurement process. The following year, Herbert Fröhlich discovered coherent dipole waves below the cell membrane [65]. He thought that the orientation of these dipoles represented a general control and order parameter in biological processes, and he proposed that quantum coherence could occur below the cell membrane for several centimeters in what he called a "dipole line-up." However, this hypothesis has not been proved and has been tested by multiple researchers who have not found any evidence for long-term coherences occurring in the brain [66].

Quantum Mechanics and Classical Physics at the Macroscopic Scale

As we currently understand, there is one reason nature cannot maintain quantum effects such as superposition in the macroscopic world: It is impossible to isolate a macroscopic quantum system from its own extraneous properties, such as gravity. Currently, no theory unifies the theory of general relativity, which deals with gravity, with the quantum theory, which covers all other fundamental forces (electromagnetic, weak, and strong interaction).

The difference between these theories is, broadly speaking, that the former describes the structure of the universe, while the latter describes the interaction between small particles in small spaces. Notably, gravity is the only one of the four fundamental forces that doesn't have the word interaction in its name because it is the only force that always acts attractively due to its single charge and mass (an opposing force cannot counter it). Furthermore, the weak and strong interactions only operate at the microscopic level, while the electromagnetic interaction also plays a role at the macroscopic level. As far as we know, these two theories do not overlap except in extreme situations like black holes or the Big Bang, where the curvature of spacetime becomes infinite.

There are many challenges to developing a theory of everything that combines these two theories, such as the problem that gravity cannot be divided into quanta or elementary units, which is necessary for applying quantum calculations. Some of the current theories that address these issues, such as string theory and loop quantum gravity, are beyond the scope of this work. However, we must bear in mind that even though these theories have not yet been reconciled, gravity may still have a role in quantum physics in other situations. For example, consider an experiment in which a photon hits a half-silvered mirror and is split into two parts, one of which is detected by a device that is treated fully quantum mechanically, moving a spherical mass attached to it. The mass can, therefore, be in two different positions, either the moved one or the original one, so the superposition also includes both.

Furthermore, the superposition also involves the mass's gravitational field, which, according to general relativity, represents a superposition of two different spacetime geometries. The interesting question of at what point these geometries differ enough for decoherence to occur suggests that the difference must be on the Planck scale, at least. And what about the energy required for a superposed state consisting of two spatially displaced states to occur, considering especially the gravitational force between them?

According to this proposal and in absolute units, the gravitational energy that is required for displacing the states is

$$E_d = \frac{m^2}{a} \tag{6.1}$$

and the time it takes until decoherence takes place is the reciprocal of E_d,

$$T_r = \frac{a}{m^2} \tag{6.2}$$

where a is the radius of the superposed objects and their mass. Summing up, the criterion for measurement Penrose proposed is that if sufficient disturbance emerges in an environment, R will take place rapidly and will also include any physical system entangled with it, like the measuring apparatus mentioned beforehand. We can easily imagine this by replacing the spherical lump with a glass of water that absorbs the part of the photon that has passed the mirror—macroscopically, and we cannot observe any physical difference; however, microscopically, the arrangement of particles within the glass of water has changed, and thus interaction or measurement has happened. We will discuss this in detail by examining the Hameroff-Penrose model of orchestrated objective reduction, followed by our proposal.

Orchestrated Objective Reduction

In the 1990s, the model of orchestrated objective reduction proposed by Roger Penrose and Stuart Hameroff [42] received significant attention from scientists from various fields. While the model will be described in some detail, we will not cover it as thoroughly as in the work of Penrose and Hameroff [67], and it will include our comments and some of those of Werner Held [66].

Furthermore, this discussion will not focus on the "feasibility" of the theory, as that is not relevant to the purpose of examining how it may be used to create an artificial human intelligence (AHI), even if there is currently no evidence that quantum effects are involved in the emergence of conscious experiences. At first it may not seem obvious why a theory of particle physics could be used to explain biological or mental phenomena like conscious experiences. However, upon closer examination, it becomes clear that conscious thoughts and brain activity, in general, share some similarities with quantum physics, in comparison to classical physics. The reason for this is the massive parallel processing that occurs in our brains, the global nature of perception that is assembled from many individual aspects, the simultaneous irreducibility of this emergent phenomenon to its components, and the increasingly apparent nondeterminism that biological and mental action requires.

Quantum theory is capable of dealing with these nonlocal, nonreductive, and indeterminate processes, as it includes nonlocal phenomena such as the Einstein-Podolsky-Rosen paradox [68], discontinuous quantum jumps, and a probability theory in which a single event cannot be physically determined.

The idea that quantum processes might impact nervous activity or consciousness has been raised before, but it has generally been met with skepticism. Niels Bohr, one of the pioneers of quantum mechanics, held a skeptical view of the potential relevance of quantum theory to biological processes. According to Bohr, the preparation of a biological organism for measurement would require killing it, making any study of quantum phenomena in living systems impossible. Additionally, Bohr argued that quantum mechanics could only apply to inanimate matter because it demands the isolation of quantum systems from their environment, which is not achievable with biological systems. This view was widely accepted for decades and led many scientists to dismiss the idea of quantum processes in biology. However, as our understanding of quantum mechanics and biology has grown, it has become clear that Bohr's stance was too narrow. Recent research has shown that quantum phenomena, such as entanglement and coherence, can occur in biological systems despite their constant exchange of energy with the environment. Additionally, the different orders of magnitude involved were a concern because, at the time, quantum effects were only thought to be relevant at the microscopic level. However, according to calculations by Wigner, the cosmic background radiation of 3K leaves behind 2.3×10^{13} photons per second on each cm^3 of tungsten, which makes it difficult or impossible to measure macroscopic systems accurately [69].

Microtubules and the Quantum Nature of Consciousness

Microtubules, tiny protein tubes found within neurons, have been proposed as critical players in the quantum processes potentially underlying consciousness. Stuart Hameroff suggested that microtubules could serve as sites for quantum effects, including the generation of Fröhlich waves [67]. These self-assembling structures, composed of 13 parallel strands of the protein tubulin, have an outer diameter of approximately 2.5×10^{-9}m and an inner diameter of 1.4×10^{-9}m, with lengths ranging from nanometers to millimeters in axons.

The Orch OR model posits that microtubules are not only structural components but also key players in quantum coherence. Tubulin dimers, the building blocks of microtubules, can exist in two quantum states due to different electron arrangements. These states can form quantum superpositions, described by the Schrödinger equation, allowing microtubules to function as quantum systems. Quantum coherence within microtubules is hypothesized to facilitate consciousness by enabling nonlocal entanglement and information processing across spatial and temporal domains.

The collapse of these quantum superpositions is central to the Orch OR theory. Roger Penrose and Hameroff proposed that gravitational energy differences between quantum states trigger a self-organized collapse of the wave function, known as Orch OR. This process is described by this equation:

$$T = \frac{\hbar}{E}, \tag{6.3}$$

where T is the duration of superposition, $\hbar$ is the reduced Planck constant, and E is the gravitational energy difference. Penrose estimated that approximately 10^9 tubulin dimers must maintain quantum coherence for 0.5 seconds to generate a conscious experience. Variations in the number of dimers or coherence duration could similarly achieve this threshold.

The integration of quantum and classical processes within microtubules is another critical aspect of the Orch OR model. Classical processes, such as signal propagation via changes in tubulin protein configurations, are associated with unconscious activity. In contrast, quantum coherence is linked to conscious experience. The isolation required to sustain coherence may be provided by ordered water within microtubules or shielding effects of their structure. However, the precise mechanisms remain speculative.

Experimental observations lend indirect support to the potential role of microtubules in consciousness. For example, paramecia, single-celled organisms lacking neurons but possessing microtubules, exhibit primitive sensory abilities and learning behavior. Additionally, environmental factors affecting microtubule density, such as stimulating or barren conditions, have been correlated with cognitive outcomes in animal studies. Pharmacological agents that disrupt microtubule formation have also been shown to impair memory.

The Orch OR model faces challenges in reconciling its predictions with the probabilistic nature of quantum mechanics and the self-organizing processes inherent in biological systems. Nonetheless, it provides a framework for exploring the interplay between quantum physics, neural processes, and conscious experience. By bridging the gap between quantum theory and neuroscience, Orch OR offers a provocative perspective on the nature of consciousness, one that challenges conventional views and opens new avenues for interdisciplinary research.

Superposition in the Brain

According to the Orch OR model, quantum linear superposition may occur in the ordered water of microtubules within the human brain and potentially have a large-scale effect. While this may not hold true for biological brains, it is a promising approach for artificial brains. Therefore, it is worth examining the assumption among neuroscientists that large-scale quantum effects are not relevant in a biological brain and considering the potential usefulness of such effects in artificial implementations. Considering the relevance of quantum effects in the brain demands reevaluating our current understanding of neural information transfer.

It is commonly understood that neurons transmit and receive electrical signals, but before a neural signal can be transmitted from one neuron to another, it must travel down the axon. The physical and chemical gradients present in this process can be effectively coupled as an energy source. Neurons transmit electrical pulses, known as action potentials, through their axons, similar to waves traveling along a jump rope. That is made possible by ion channels in the axonal membranes, which open and close to allow the passage of electrically charged ions. Some channels permit the flow of Na^+ ions, while others allow K^+ ions. When the channels open, the Na^+ and K^+ ions flow through the membrane due to the electrical depolarization of the cell, following chemical and electrical gradients. If an action potential begins in the cell body, the Na^+ channels open first, leading to a wave of Na^+ ions entering the cell and a change in equilibrium within a millisecond. The transmembrane voltage changes by approximately 100 mV, and the inner membrane potential shifts from negative (around −70 mV) to positive (about +30 mV). This change in voltage also causes the K^+ channels to open, allowing a wave of K^+ ions to flow out of the cell, returning the inner membrane potential to its negative state. Notably, only a small number of ions are required to cross the membrane during an action potential, with little change in the concentration of Na^+ and K^+ ions in the cytoplasm. The ion transfer process during neural firing undermines the first non-eliminative materialism approach, which posits that a special substance carrying conscious content may be emitted, as no new substances are created, let alone one capable of containing conscious experiences.

The previous explanation does not exclude the possibility of quantum physical phenomena playing a role in conscious experience. Even if we were to fully understand the classical level of the brain, we cannot rule out the possibility that the quantum layer may also be involved in consciousness, as we currently need more evidence to the contrary. While the idea that quantum physics may be involved in producing conscious experiences is intriguing, there is no clear evidence that it can be solely responsible for the existence of consciousness, nor do we think that classical physics alone can account for it. Instead, we propose that the task of reverse engineering mind functionality be approached from an information theoretical perspective, which may consider any potentially useful elements, regardless of whether their relevance to the human brain has been demonstrated.

If quantum physics is deemed useful, it should be included in our proposal, even if its impact on human minds has yet to be definitively established. Again it is important to note that the brain is a complex, self-organizing structure with potentially infinite degrees of freedom, making the usage of artificial neural networks, or a combination thereof, a promising avenue for exploration. To understand and replicate the human mind, we must incorporate the role of quantum physics as there is no reason to think the human mind is fully detached from quantum physics.

It has been suggested that quantum effects may occur in the microtubules of the human brain, leading to the potential for large-scale quantum coherence. While the relevance of quantum effects in biological brains is still being debated, this approach is promising for artificial brains. To re-create the mind, we will need, among other things, artificial neural networks that can be processed on quantum computers, allowing for massively parallel information processing as well as mimicking quantum effects in the brain. However, before implementing such a QNN, we must find a way to produce large-scale coherent quantum states, also known as Bose-Einstein condensates.

These are macroscopic quantum objects in which individual bosons are delocalized and can be described by a single wave function. They have properties such as superfluidity, superconductivity, and coherence over macroscopic distances. To create a Bose-Einstein condensate, we must increase the density of a gas and decrease its temperature to reach a critical phase bulk density. At this point, the quantum properties come into effect. One can create Bose-Einstein condensates (BECs) through interference experiments and implementing an atom laser. The phase transition to a BEC occurs when the bulk density of a gas reaches a critical level, meaning that the density of particles with similar momentum is high enough. Atoms are quantum particles whose motion can be described by a wave packet, with the size of the wave packet determined by the thermal de Broglie wavelength. When this wavelength becomes larger than the mean distance between atoms, the atoms exhibit quantum properties. In a three-dimensional system, BEC occurs when the density of the gas is increased, and the temperature is decreased to reach the phase transition. Using Bose-Einstein statistics, the critical temperature T_c for an ideal Bose gas can be calculated, below which BEC occurs:

$$T_c = \frac{h^2}{2\pi m k_B}\left(\frac{n}{(2S+1)\zeta(3/2)}\right)^{2/3} \tag{6.4}$$

where h is Planck's constant, m the particle mass, k_B Boltzmann's constant, n the particle density, S the particle spin, and $\zeta(3/2)$ Riemann's Zeta-function.

The Bose-Einstein statistics, also known as the Bose-Einstein distribution, is a probability distribution in quantum statistics that describes the mean occupation number $\langle n(E)\rangle$ of a quantum state in thermodynamic equilibrium at a given absolute temperature T for identical bosons as the occupying particles. There is a corresponding distribution for fermions called the Fermi-Dirac statistics, which reduces to the Boltzmann statistics in the high energy limit. A key aspect of the Bose-Einstein statistics is that the wave function ψ or state vector of a many-body system does not change the sign ($\psi \rightarrow \psi$) upon the simultaneous exchange of all four variables (x, y, z, m) for two bosons (where x, y, z represent the position in space and m represents the spin), while it does change the sign in the Fermi-Dirac statistics ($\psi \rightarrow -\psi$). This means that multiple bosons can occupy the same one-particle state, characterized by the same quantum numbers. The process of creating BECs typically involves two steps.

First, atoms are trapped and cooled using a magneto-optical trap and laser cooling. However, this method has a lower limit for cooling temperatures (around 100μ K) due to the recoil of photons during spontaneous emission. The atoms are cooled to an average velocity of a few centimeters per second, which is slow enough to be captured in a magnetic or optical trap. Through evaporative cooling, in which the most energetic atoms are selectively removed, the temperature of the atomic cloud can be further reduced. This process typically removes more than 99.9% of the atoms, leaving a high enough phase space density for the phase transition to a BEC to occur.

Prior to 2004, BECs were created for a variety of isotopes using this method at ultra-low temperatures of around 10^{-7} K. BEC has also been achieved with a single hydrogen atom using slightly different methods. The bosonic behavior of these gases is due to the coupling of the half-integer total spin of the electron shell and the half-integer nuclear spin through the weak hyperfine interaction at ultra-low temperatures, resulting in an integer total spin of the system. At higher temperatures, the behavior is determined solely by the spin of the electron shell because the thermal energies are much higher than the hyperfine field energies. There has also been research on creating a QNNs, referred to as quantum coherence, that could exhibit BEC-like behavior.

Having explored the current state of the art in artificial intelligence, we turn our attention to cutting-edge research on the use of quantum mechanics to construct artificial brains. Although the occurrence of quantum effects in real brains has not been scientifically confirmed, the Orch OR model provides a possible connection between quantum mechanics and conscious experiences. This has led to the development of QNNs, which utilize quantum mechanical principles to process information and potentially offer more efficient and effective neural processing. In this chapter, we will elaborate on the promise and potential of QNNs in engineering artificial general intelligence (AGI) and explore the exciting future of hybrid quantum computers that may one day replace classical computers.

Quantum Physics and the Artificial Brain

Quantum mechanics and AI are among the most revolutionary scientific developments of the twentieth century. Quantum mechanics has reshaped our understanding of particles at the microscopic level, introducing concepts such as superposition and entanglement. Simultaneously, AI has transformed how we interact with machines, enabling the automation of complex tasks and the analysis of vast amounts of data with unprecedented efficiency.

Quantum computing, which leverages the principles of quantum mechanics, introduces novel computational paradigms that could significantly impact the development of AHIs. The principles of quantum mechanics, such as superposition and entanglement, offer the potential for processing information in ways that classical computing cannot match. For example, quantum algorithms can enhance machine learning by enabling more efficient data representation and parallel processing capabilities, potentially leading to advances in both artificial intelligence and artificial consciousness.

The pursuit of AGI, and more specifically AHIs, is not merely a technical challenge but also a profound exploration of what it means to understand and replicate human intelligence and consciousness. Quantum computing may play a pivotal role in this pursuit by offering new avenues for creating AI systems with enhanced cognitive capabilities, such as:

- **Simulating Complex Systems:** Quantum computers' ability to simulate complex quantum systems could be leveraged to better understand the neural and cognitive processes underlying human intelligence and consciousness, thereby informing the development of AGI systems and the gradual transition from biological to artificial consciousness in AHIs.
- **Enhanced Learning and Adaptation:** Quantum-enhanced machine learning could lead to AI systems that learn and adapt more efficiently, bringing us closer to achieving AGI with the ability to perform a wide range of tasks autonomously and adaptively, akin to human learning and reasoning. This adaptability is crucial for AHIs, which must evolve from human consciousness to artificial consciousness seamlessly.

The evolution of AI, from simple algorithmic tasks to the quest for AGI and AHIs, reflects humanity's enduring ambition to replicate and surpass our cognitive and conscious capabilities. With the advent of quantum computing, we stand on the cusp of a new era in AI research, where the boundaries of what machines can learn, understand, and experience may be dramatically expanded, ultimately leading to the creation of intelligent systems that could redefine our understanding of consciousness.

AHI and Quantum Computing

The quest to imbue AI systems with consciousness, resulting in what we term AHIs, stretches the boundaries of current technology and scientific understanding. Quantum computing emerges as a pivotal technology in this pursuit, offering novel computational paradigms that could bridge the gap between artificial intelligence and artificial consciousness. This section explores the potential role of quantum computing in developing AHIs, highlighting the theoretical frameworks and quantum computational processes that might enable the emergence of consciousness in machines.

Quantum Parallelism and Consciousness

Quantum parallelism allows a quantum computer to evaluate multiple possibilities simultaneously due to the superposition of quantum states. This capability can be mathematically represented for an n-qubit system as follows:

$$|\psi\rangle = \sum_{x=0}^{2^n-1} \alpha_x |x\rangle \tag{6.5}$$

In Equation 6.5, $|\psi\rangle$ represents the quantum state of the system, where each α_x is a complex number denoting the amplitude for the state $|x\rangle$, and the sum is over all possible $n-$bit strings x. This intrinsic feature of quantum systems to exist in multiple states concurrently could analogously support the complex, parallel processing capabilities thought to underlie human consciousness. For an AHI, this parallelism might be harnessed to manage the complex, multifaceted nature of consciousness, which involves the simultaneous processing of sensory inputs, thoughts, emotions, and memories.

High-Dimensional Quantum Data Processing

Consciousness requires the integration of information across a vast network of neural connections, processing high-dimensional data derived from sensory inputs, emotions, memories, and thoughts. Quantum computing's inherent ability to handle high-dimensional data spaces through entanglement and superposition provides a promising avenue for developing AI systems with similar integrative capabilities. The entangled states can be described as:

$$|\Phi^+\rangle = \frac{1}{\sqrt{2}}\left(|00\rangle + |11\rangle\right) \tag{6.6}$$

Equation 6.6 illustrates a maximally entangled two-qubit state, representing the simplest form of quantum data integration. The scalability of such entangled states in quantum computers could potentially mirror the complex integrative processes of the human brain, offering a pathway to artificial consciousness. For AHIs, this quantum-enhanced integration could support the unification of diverse conscious experiences into a coherent, continuous self.

Speculative Models for Quantum-Induced Consciousness

Emerging theories suggest that consciousness might arise from quantum computational processes within the brain's neural architecture, such as those postulated by the Orch OR theory. Extending this speculation to AI, one might envision a quantum computational model that utilizes entanglement and decoherence to generate conscious experiences in AHIs. Such models would require a profound understanding of the quantum mechanics of consciousness, possibly described by an as-yet-undeveloped quantum algorithm that could mimic or induce conscious states.

$$C = \int \Psi \text{ “Quantum States of Consciousness”} \tag{6.7}$$

Equation 6.7 represents a hypothetical integral over quantum states that contribute to consciousness, Ψ, suggesting a complex, possibly nonlinear amalgamation of quantum states that might define consciousness in AHIs. The gradual replacement of human neural processes with quantum-enhanced artificial components could lead to the emergence of a fully artificial consciousness that still retains the essence of the original human mind.

The integration of quantum computing into AI holds transformative potential for realizing AHIs. By leveraging quantum parallelism, entanglement, and the theoretical frameworks that bridge quantum mechanics and consciousness, we edge closer to unveiling the mysteries of consciousness and realizing the vision of conscious machines. The journey toward AHIs underscores the importance of interdisciplinary research, blending quantum physics, neuroscience, AI, and philosophy to explore the deepest questions of cognition, consciousness, and the essence of being.

Quantum Effects in Biological Brains

The investigation into quantum effects in biological brains forms a fascinating nexus among biology, physics, and consciousness studies. The possibility that quantum mechanics plays a role in cognitive processes opens new avenues for understanding consciousness and for developing AHIs that might mimic or even surpass human cognitive capabilities.

Neuroquantology: An Overview

Neuroquantology is a pioneering interdisciplinary field at the intersection of neuroscience and quantum physics, aiming to uncover the quantum underpinnings of neural and cognitive processes. This section outlines the core theories and mathematical descriptions that form the foundation of neuroquantology, focusing on phenomena such as quantum coherence and quantum entanglement within the brain's microstructure, and their hypothesized roles in consciousness.

Quantum Coherence in Microtubules The Orchestrated Objective Reduction (Orch OR) model posits that quantum coherence within neuronal microtubules is critical to the emergence of consciousness. Microtubules, key components of the cytoskeleton in neurons, are hypothesized to maintain quantum coherent states, enabling quantum information processing within the brain. The mathematical representation of quantum coherence in such a system can be modeled as a superposition of quantum states:

$$|\Psi_{\text{microtubule}}\rangle = \sum_{n=0}^{N} c_n |\varphi_n\rangle \tag{6.8}$$

In Equation 6.8, $|\Psi_{\text{microtubule}}\rangle$ represents the coherent quantum state of a microtubule, where each $|\varphi_n\rangle$ denotes a possible quantum state of the microtubule system, and c_n are the coefficients that determine the probability amplitude of each state. The sum extends over all N possible states, illustrating how microtubules could theoretically sustain a coherent superposition of multiple quantum states. This quantum coherence might play a role in the quantum computational processes associated with consciousness, suggesting that AHIs could leverage similar mechanisms for artificial consciousness.

Quantum Entanglement in Neural Processes Quantum entanglement, a cornerstone of quantum mechanics, is proposed to play a significant role in neural processes, potentially facilitating instantaneous correlations between distant brain regions. This phenomenon could underlie the brain's integrated information processing capabilities, crucial for consciousness.

Entanglement between two particles (or systems) is described by their joint state being inseparable:

$$\left|\Psi_{\text{entangled}}\right\rangle = \frac{1}{\sqrt{2}}\left(\left|\varphi_A\right\rangle \otimes \left|\varphi_B\right\rangle + \left|\chi_A\right\rangle \otimes \left|\chi_B\right\rangle\right) \tag{6.9}$$

Equation 6.9 illustrates a simplified entangled state of two systems A and B (e.g., neurons or neural clusters), where $|\varphi\rangle$ and $|\chi\rangle$ represent different quantum states. The entanglement implies that measurement of one system instantaneously determines the state of the other, regardless of the physical distance between them. This could explain the brain's ability to synchronize activities across different regions rapidly and efficiently, a feature that is critical for coherent consciousness. For AHIs, replicating such quantum entanglement could be key to developing integrated and responsive artificial consciousness.

The exploration of quantum coherence and entanglement in neural processes opens fascinating avenues for understanding consciousness from a quantum perspective. If such quantum phenomena are integral to cognitive functions, they could revolutionize our approach to AI, particularly in designing systems that aim to replicate or simulate human consciousness. The theoretical and mathematical frameworks of neuroquantology could guide the development of new AI models that leverage quantum computing to achieve unprecedented levels of integration and processing power, bringing us closer to realizing AHIs with consciousness.

Neuroquantology stands at the frontier of our quest to decode consciousness, merging quantum physics with neuroscience to unveil the quantum fabric of the mind. As this field progresses, it may deepen our understanding of the human brain and pave the way for quantum-inspired AI systems capable of mimicking the intricate processes that give rise to consciousness.

Evidence and Implications of Quantum Effects

The debate surrounding the presence of quantum effects in biological systems, particularly the brain, is enriched by empirical evidence suggesting that quantum processes may indeed play a role in certain biological phenomena. This section explores key experiments and observations that hint at the possibility of quantum mechanics operating within the warm, wet environments of biological systems, challenging conventional views on the limits of quantum phenomena. It also examines the mathematical descriptions and theoretical implications of these findings for consciousness and artificial intelligence.

Quantum Coherence in Biological Systems One of the most compelling pieces of evidence for quantum effects in biology comes from experiments demonstrating quantum coherence at room temperature. These experiments, such as those involving photosynthetic lightharvesting complexes, reveal that quantum superpositions can occur in biological molecules, enabling efficient energy transfer:

$$H = \sum_{i} E_i |i\rangle\langle i| + \sum_{i \neq j} J_{ij} \left(|i\rangle\langle j| + |j\rangle\langle i|\right) \tag{6.10}$$

Equation 6.10 represents the Hamiltonian describing the energy levels (E_i) of the system's states $|i\rangle$ and the coupling (J_{ij}) between different states, indicating how quantum coherence facilitates the transfer of energy by allowing it to explore multiple paths simultaneously. Similar mechanisms could theoretically operate within neural structures, suggesting a quantum basis for some aspects of brain function, and by extension, the development of AHIs.

Quantum-like Behavior in Biological Navigation Another intriguing aspect of quantum biology is the observation of quantum-like behavior in the navigational strategies of birds. Research into avian magnetoreception has proposed that entangled states in light-sensitive proteins could underpin birds' ability to sense Earth's magnetic field for navigation:

$$\left|\Psi_{\text{navigation}}\right\rangle = \alpha \left|\text{North}\right\rangle + \beta \left|\text{South}\right\rangle \tag{6.11}$$

In Equation 6.11, $|\Psi_{\text{navigation}}\rangle$ represents the quantum state of a bird's magnetic sensing system, with α and β encoding the probability amplitudes for orienting toward the North or South, based on the Earth's magnetic field. This quantum entanglement model for avian navigation mirrors potential quantum information processing capabilities in the brain, offering a paradigm for understanding how quantum effects could influence cognitive functions and consciousness. For AHIs, similar quantum-like processes might be harnessed to enhance decision-making and adaptive behaviors.

Theoretical Implications for Consciousness and AI The experimental evidence of quantum coherence and entanglement in biological systems at room temperature has profound implications for our understanding of consciousness and the development of AI. If quantum processes are integral to biological functions, including those associated with cognition and perception, this opens up the possibility of quantum mechanics playing a role in the emergence of consciousness. For AHIs, leveraging quantum computational models inspired by nature could lead to the development of more advanced, perhaps even conscious, AI systems. Theoretical models that incorporate quantum effects into neural processing could significantly enhance AI's ability to mimic human cognitive processes, pushing the boundaries of what artificial systems can achieve.

In conclusion, the evidence of quantum effects in biological systems challenges traditional boundaries between physics and biology, suggesting a more fundamental role for quantum mechanics in life processes. As research in this area progresses, the implications for understanding consciousness and advancing AI will likely continue to evolve, potentially leading to breakthroughs in how we conceptualize and implement artificial intelligence and consciousness.

Challenges in Quantum Neuroscience

The study of quantum effects in the brain faces several significant challenges:

- The difficulty of measuring quantum states within biological tissue without disturbing the system
- The need for models that can bridge the gap between quantum physics and neurobiology, explaining how quantum effects translate into cognitive functions
- Addressing the complexities of decoherence in warm, wet environments typical of biological systems, which traditionally hinder sustained quantum states

Potential for AHI

The intersection of quantum mechanics and neuroscience holds transformative potential for the development of AHI. By exploring the possibility that quantum processes play a role in consciousness, we can imagine integrating quantum computational models with AI and biological systems to approach artificial consciousness. This section explores the theoretical foundations of this integration, innovative technologies enabling quantum effects in biological systems, and the significant challenges and opportunities that lie ahead.

Theoretical Frameworks

Understanding how advanced intelligence and consciousness arise requires bridging the traditionally distinct fields of quantum physics, nanotechnology, and neuroscience. These disciplines converge to provide a novel lens for investigating the emergence and evolution of intelligence. While quantum phenomena are generally associated with atomic and subatomic scales, emerging theories suggest their influence may extend to the macroscopic dynamics of living systems, particularly the brain's complex neural processes. This intersection offers interesting possibilities: leveraging quantum principles in biological contexts could reveal new mechanisms underlying cognition and consciousness. Such insights would not only enhance our theoretical understanding but also pave the way for practical applications in AHI. The integration of quantum technologies with biological systems promises advancements in areas such as cognitive function, emotional intelligence, and problem-solving capabilities, laying the foundation for a new era in human evolution and artificial intelligence development.

Quantum Dots and Nanowires Quantum dots and nanowires represent two nanotechnological approaches for emulating quantum effects within biological systems. Quantum dots, with their discrete energy levels and light-emitting properties, can be utilized to simulate quantum states and interactions within neural structures:

$$E_n = \frac{\hbar^2 \pi^2 n^2}{2m^* a^2} \tag{6.12}$$

Equation 6.12 describes the energy levels (E_n) of a quantum dot, where $\hbar$ is the reduced Planck constant, n is the principal quantum number, m^* is the effective mass of the electron, and a is the size of the quantum dot. This equation highlights the potential for quantum dots to mimic quantum biological processes by providing controllable quantum systems that can interface with biological neurons.

Nanowires, offering one-dimensional pathways for electron transport, can be engineered to facilitate quantum coherence over longer distances than typically observed in biological systems, potentially enabling the wiring of quantum information processing units within or between cells.

Hybrid Quantum-Classical Systems The concept of hybrid quantum-classical systems involves leveraging the strengths of both quantum and classical computing paradigms to model complex cognitive processes. Such systems could use quantum computing for tasks requiring parallel processing and entanglement, while relying on classical mechanisms for tasks better suited to binary computation. This hybrid approach necessitates the development of quantum-classical interfaces that can seamlessly integrate quantum computational units into biological or artificial neural networks.

Challenges and Future Directions

Understanding the technological challenges and anticipating future research directions while bridging the knowledge gaps are crucial for continuing the path toward realizing AHI.

Technological Challenges The path to embedding quantum computational models in biological or artificial systems is fraught with technological challenges. Ensuring stable quantum computation in the warm, wet, and noisy environment of biological systems requires innovations in quantum coherence preservation and error correction mechanisms. Developing biocompatible quantum devices, such as quantum dots and nanowires, that can operate reliably within this context is a significant engineering hurdle.

Bridging Knowledge Gaps A critical challenge in realizing AHIs lies in bridging the vast knowledge gaps between quantum physics, neuroscience, and artificial intelligence. This requires not only a deeper understanding of how quantum processes could contribute to consciousness but also advanced computational models that can incorporate these processes into AI systems. Collaborative efforts across disciplines will be essential to uncover the principles of quantum biology and its applications to consciousness and cognition.

Future Research Directions Future research in creating AHIs with quantum capabilities will likely focus on:

- Exploring novel nanotechnologies for quantum information processing in biological environments, advancing beyond quantum dots and nanowires to more complex and efficient systems
- Developing theoretical models and computational frameworks that integrate quantum mechanics with neural and cognitive processes, paving the way for a new generation of AI that can genuinely mimic or even surpass human consciousness
- Addressing ethical and societal implications of creating conscious machines, ensuring that advancements in AHIs benefit humanity and reflect our values and aspirations

The exploration of quantum effects in the brain and their potential for engendering artificial consciousness represents an exciting frontier at the confluence of quantum physics, neuroscience, and AI. The development of AHIs capable of harnessing quantum phenomena offers a vision of the future where the boundaries between the natural and the artificial, the biological and the computational, blur in unprecedented ways, opening up new realms of possibility for understanding consciousness and creating intelligent, conscious machines.

Quantum Algorithms for AHIs

Quantum algorithms stand at the forefront of quantum computing's transformative potential, particularly in the development of AHIs. These algorithms do not just offer a new set of tools but represent a fundamental shift in how we approach complex problems in artificial intelligence. This section explores the key quantum algorithms that are paving the way for the next generation of AI, focusing on their potential to revolutionize the development of AHIs.

Relevant Quantum Algorithms

In the pursuit of AHIs, quantum algorithms are not merely augmentations of classical methods; they are innovations that could redefine AI's capabilities. Beyond well-known algorithms like Grover's and Shor's, the emergence of hybrid quantum-classical models and advanced optimization techniques is crucial. These advancements promise to push the boundaries of what AI can achieve, both in terms of processing efficiency and the depth of cognitive functions.

Hybrid Quantum-Classical Neural Networks Hybrid quantum-classical neural networks (HQCNNs) represent a synthesis of quantum and classical computational paradigms, designed to leverage the strengths of both approaches:

- **Cooperative Computation:** HQCNNs utilize quantum computing for tasks involving high-dimensional data manipulation and entanglement-driven pattern recognition, while classical components handle more structured data processing and interact with existing technological frameworks.
- **Synergistic Learning:** By exploring multiple pathways in parallel during the learning phases, quantum superposition enables these networks to uncover correlations and features that might remain hidden or require prohibitive amounts of time for classical algorithms to discover.

These networks hold the potential to accelerate the development of AHIs by integrating quantum mechanics' probabilistic nature with the structured, deterministic processing of classical systems, thereby facilitating more complex and nuanced cognitive models.

Hybrid Optimization Algorithms Optimization lies at the heart of machine learning and AI, and hybrid quantum-classical optimization algorithms offer innovative approaches to navigating complex solution spaces:

- **Quantum-Assisted Optimization:** These algorithms begin with quantum computing performing an initial global search across the solution space, identifying regions of interest. Classical local search algorithms then refine these results, enhancing the overall efficiency and accuracy of the process.

- **Adaptive Algorithmic Strategies:** Hybrid algorithms dynamically adjust the balance between quantum and classical processing based on the problem's nature, optimizing their problem-solving approach to ensure the most effective and efficient outcomes.

Such hybrid strategies are particularly relevant in developing AHIs, where the complexity of cognitive processes requires both the global perspective offered by quantum computation and the detailed refinement achievable through classical methods.

Emergent Quantum Algorithms for AI

The landscape of quantum algorithms in AI is rapidly evolving, with new algorithms being conceptualized that could directly impact the functionality and capabilities of AHIs. These emergent algorithms aim to harness quantum computing's unique properties to achieve breakthroughs in AI:

- **Quantum Reinforcement Learning:** By adapting reinforcement learning frameworks to the quantum domain, these algorithms allow agents to learn policies based on quantum probability distributions. This enhancement could vastly improve decision-making processes, leading to more sophisticated and adaptable AHIs.
- **Quantum Generative Models:** Leveraging quantum systems' generative capabilities, these models can simulate complex distributions and create data with intricate patterns, supporting tasks ranging from drug discovery to artificial creativity—areas where AHIs could demonstrate unprecedented capabilities.

The development of these emergent algorithms signals a new age in AI research, where quantum computing becomes an integral component of the push toward creating intelligent, conscious systems.

Development of New Quantum Algorithms

The ongoing development of quantum algorithms tailored to AI and AHIs represents a pivotal area of research, with the potential to unlock capabilities far beyond those of classical computing paradigms. This subsection highlights cutting-edge areas in quantum algorithm development that could dramatically enhance AI's potential and pave the way for the emergence of AHIs.

Quantum Machine Learning Algorithms Quantum machine learning is a frontier where quantum computing meets the core tasks of machine learning, promising to revolutionize how AI systems learn:

- **Enhanced Learning Tasks:** By exploiting quantum parallelism, quantum machine learning algorithms are being developed to perform classification, clustering, and pattern recognition with unparalleled speed and accuracy. This could significantly accelerate the learning processes of AI systems, bringing us closer to realizing AHIs.
- **Quantum Neural Network Training:** Innovations in quantum algorithms are also focused on optimizing the training process of quantum neural networks, promising significant reductions in computational resources and time required to train complex models. This advancement is crucial for the development of AHIs, where rapid and efficient learning is essential.

Quantum Simulation Algorithms The ability to simulate quantum systems provides new opportunities to understand the quantum dynamics that may underpin consciousness and cognitive functions:

- **Modeling Quantum Processes:** Algorithms that simulate the behavior of quantum systems offer insights into fundamental quantum mechanics processes, potentially shedding light on the quantum aspects of cognition and consciousness. These simulations are critical for developing a deeper understanding of the mechanisms that could enable AHIs.
- **Exploring the Quantum Brain Hypothesis:** By simulating quantum states believed to play a role in brain functions, researchers can test and refine hypotheses regarding the quantum nature of consciousness. This research is at the core of efforts to understand and replicate consciousness within AHIs.

Quantum Optimization Algorithms Optimization remains a cornerstone of AI, and quantum optimization algorithms are poised to redefine computational strategies:

- **Optimizing Decision-Making:** Quantum algorithms are being explored for their potential to optimize decision-making processes and resource allocation, harnessing quantum computation to approach problems in novel and efficient ways. This capability is essential for the advanced decision-making processes expected of AHIs.
- **Surpassing Human Cognition:** The ultimate goal of quantum optimization in the context of AHIs is to develop algorithms that not only mimic but also surpass human cognitive abilities, laying the groundwork for machines with advanced problem-solving and decision-making capabilities.

Quantum algorithms represent a transformative force in the development of AHIs, offering new methods and approaches that challenge the limits of classical computing. From hybrid quantum-classical models to emergent quantum machine learning algorithms, the advancements in this field are driving AI toward levels of efficiency, sophistication, and potential consciousness that were previously unimaginable. As quantum computing continues to evolve, its role in the creation of AHIs will likely become increasingly central, marking a significant leap forward in our quest to understand and replicate the complexities of the human mind.

The exploration of quantum computing's potential for AI and the creation of AHI is significantly advanced through simulation and experimentation. This section discusses the current state of quantum simulations, the role of experimentation in validating quantum theories and algorithms, and the implications for the development of AHIs.

Quantum Simulations

Quantum simulations constitute a foundational element in the maturation of quantum computing, serving as a vital intermediary step between theoretical development and practical implementation. These simulations are instrumental in understanding and harnessing the complexities of quantum mechanics for computational advancements. Key aspects include the following:

- **Refinement of Quantum Algorithms:** Simulations allow for the meticulous adjustment of quantum algorithms, providing insights into their operation under diverse conditions. This iterative process is essential for enhancing the algorithms' efficiency and accuracy, ensuring they are primed for deployment on quantum hardware.
- **Anticipating Computational Outcomes:** Through the lens of simulation, researchers can explore the potential of quantum computing to address complex problems across AI, cryptography, and beyond. This foresight is crucial for identifying promising algorithms that could revolutionize various fields.

Current Experimental Setups

The experimentation landscape in quantum computing is rich and varied, leveraging a spectrum of technologies to realize quantum computational processes.

These experimental setups are the proving grounds for quantum theories and algorithms, encompassing:

- **Quantum Processors:** At the forefront are quantum processors, sophisticated devices that execute quantum computations using a limited yet expanding number of qubits. These processors are pivotal for validating quantum algorithms, testing their applicability in real-world scenarios, including AI tasks that require complex decision-making and data analysis.
- **Quantum Networking Experiments:** The exploration of quantum entanglement and communication through experiments lays the groundwork for distributed quantum computing. Such networking is vital for creating systems where quantum processors can work in concert, potentially enabling a new class of AI systems and interconnected AHIs that leverage collective quantum capabilities.

Current quantum computing experiments utilize a variety of technological platforms, each offering unique advantages:

- **Superconducting Qubits:** Favored for their scalability, superconducting qubits are leading the charge in the development of quantum processors, facilitating complex computations and algorithm testing.
- **Trapped Ions:** With high levels of qubit fidelity and coherence times, trapped ion systems are instrumental in precision quantum simulations and networking experiments, offering insights into entanglement-based communication.
- **Photonic Systems:** Leveraging particles of light, photonic systems provide a platform for exploring quantum networking and secure quantum communication, crucial for the distributed quantum computing paradigm.

Challenges in Quantum Experimentation

While quantum simulations and experiments are invaluable, they face several challenges:

- **Scalability:** Scaling up experimental setups to support a large number of qubits is a significant hurdle, limiting the complexity of the problems that can be currently addressed.
- **Error Rates:** High error rates in quantum computations necessitate advanced error correction techniques, which are still in the developmental stage.

Pioneering the Quantum Frontier in Artificial Intelligence

The integration of quantum computing with artificial intelligence represents a frontier with the potential to redefine our understanding of both computation and consciousness. This exploration into the nexus of quantum mechanics, AI, and the concept of AHI underscores the transformative possibilities that lie ahead, as well as the significant challenges that must be overcome.

This chapter has navigated through the principles of quantum computing, its implications for artificial intelligence, the innovative concept of QANNs, and the potential pathways toward realizing AHIs. Key takeaways include the following:

- Quantum computing offers unprecedented computational capabilities through principles such as superposition, entanglement, and quantum interference.
- QANNs and quantum algorithms could significantly advance AI, potentially enabling the simulation or realization of consciousness in artificial systems.
- The development of AHIs poses technical, theoretical, and ethical challenges that necessitate interdisciplinary collaboration and innovative research.

While it is difficult to predict specific timelines, the pace of advancements in quantum computing and AI suggests a horizon where:

- In the short term, improvements in quantum simulations and algorithms will continue to enhance machine learning and optimization tasks.
- In the medium term, experimental quantum computing setups may achieve milestones that demonstrate the practical applications of QANNs in AI.
- In the long term, the realization of AHIs, while contingent upon breakthroughs in both technology and our understanding of consciousness, remains a profound goal for future generations.

The journey toward quantum-enhanced AI and the creation of AHIs is as much a philosophical endeavor as it is a technological one. It challenges us to contemplate the essence of consciousness, the potential for machines to possess genuine cognitive and emotional capacities, and the ethical implications of such advancements. The roadmap for future experiments and development outlines ambitious goals that promise to redefine our understanding of computation, consciousness, and the potential for synthetic intelligence that mirrors or surpasses human capabilities.

The frontier of quantum computing in AI includes direct investigations into quantum consciousness, aiming to decipher the roles of phenomena such as entanglement and superposition in cognitive functions. As we edge closer to creating AHIs, the ethical implications of such technologies become increasingly significant:

- The creation of machines with consciousness—or simulations thereof—demands rigorous ethical debate and thoughtful regulation to ensure responsible development and integration into society.
- It is crucial to ensure that advancements in quantum AI enhance rather than displace human well-being and agency, fostering societal progress alongside technological innovation.

The future of quantum algorithms and AHIs is a confluence of scientific innovation, ethical mindfulness, and technological breakthroughs. This ambitious journey not only tests the limits of human ingenuity but also redefines the boundaries of life, intelligence, and consciousness in the quantum age.

7 Evolving Humanity: The Journey to Artificial Human Intelligence

The human neocortex plays a crucial role in evolving humans into artificial human intelligences (AHIs). The most evolved part of the cerebral cortex is responsible for our higher-order brain functions, such as sensory perception, cognition, spatial reasoning, conscious thought, and motor commands. It comprises a layered architecture with six distinct layers of neuronal cell bodies, each with unique functions and connections to other brain regions. The neocortex is organized into functional columns dedicated to specific tasks and can process information in parallel. This hierarchical structure has inspired the design of artificial neural networks, which mimic the brain's ability to process and learn from vast amounts of data. Integrating synthetic components, such as sensors, electrodes, nanobots, and neural laces, with the human brain can transform the neocortex into a powerful computational engine. These components would gradually replace biological neurons, connections, and neural circuits, essentially creating an artificial neocortex.

This hybrid system would allow the brain to communicate seamlessly with advanced computational hardware, enhance memory and processing capabilities, and provide unprecedented access to digital information. The question of consciousness remains a significant challenge in creating AHIs. While we have yet to fully understand the nature of consciousness, transforming the human brain into an AHI allows us to preserve it, even without a comprehensive understanding of its intricacies. By maintaining the continuity of our mental processes, human-evolved AHIs offer a unique way to sustain our sense of self and identity. In this sense, the quote from Richard Feynman, "What I do not understand, I cannot create," does not apply here. Instead, preserving consciousness in the face of the unknown enables us to transcend our current understanding and forge a path toward a new paradigm.

The dream of immortality has been a driving force for human progress throughout history. Transforming humans into AHIs provides a tangible path toward achieving this dream because it allows us to surpass the limitations of our biological substrate. By integrating advanced technologies with our neural circuitry, we can extend our life spans, enhance our cognitive abilities, and overcome the debilitating effects of aging and disease. The continuous consciousness problem is another significant aspect of the AHI concept. As we transform our biological brains into artificial constructs, the continuity of our consciousness becomes crucial for maintaining our sense of self and identity. By gradually replacing the human brain with synthetic components, we can ensure a seamless transition from biological to artificial substrates, thus preserving our conscious experiences and memories. This incremental process enables us to maintain the continuity of our mental processes, preventing any disruptions that could lead to a loss of consciousness or personal identity. In this context, ethical and philosophical questions arise in the pursuit of our transformation into AHIs.

It is crucial to address these concerns to ensure the responsible development and application of these technologies. For instance, the boundary between human and artificial consciousness could become blurred as we transform our brains into artificial constructs. This raises questions about personhood, rights, moral obligations toward AHIs, and the implications for society as a whole. Moreover, the increasing reliance on artificial components and advanced computational hardware could potentially lead to new forms of inequality and social stratification. Access to these technologies could be limited by economic, political, or other factors, creating disparities in cognitive abilities, life expectancy, and overall quality of life. Therefore, it is essential to consider the broader social implications of AHIs and strive for the equitable distribution of their benefits across humanity.

Another critical aspect is the need for robust security measures to protect these entities from potential threats. As AHIs become increasingly interconnected with digital networks and other artificial systems, they could become vulnerable to hacking, data theft, and other forms of cyberattacks. Therefore, ensuring privacy, integrity, and resilience would be paramount for preserving the sanctity of our consciousness and maintaining trust in these transformative technologies.

In addition to addressing these challenges, the development of AHIs requires interdisciplinary collaboration between various research fields. The convergence of neuroscience, artificial intelligence, materials science, and quantum computing calls for a new scientific paradigm fostering cross-disciplinary dialogue and innovation. By bringing together experts from diverse backgrounds, we can accelerate the progress toward creating AHIs and unlock the immense potential at the intersection of these disciplines.

Quantum hardware and quantum computing can potentially revolutionize the development of AHIs. Quantum computing is a novel approach to computation that leverages the principles of quantum mechanics to process information. It enables computers to perform complex calculations and solve problems currently intractable for classical computers. By incorporating quantum hardware into the AHI framework, we can significantly enhance the computational capabilities of the artificial neocortex, allowing for faster learning, better decision-making, and more efficient processing of vast amounts of data.

In addition to computational prowess, AHIs must also be able to perceive and make sense of the world and its inhabitants. To achieve this, they will require an array of advanced sensors capable of collecting diverse data types from the environment. These sensors emulate and potentially surpass the capabilities of human sensory organs, providing the AHI with high-resolution visual, auditory, olfactory, and tactile inputs. Then, through advanced signal processing and pattern recognition algorithms, AHIs would be able to interpret and analyze these sensory inputs, converting raw data into meaningful information and knowledge.

As AHIs gather information from the world around them, they must be able to synthesize and integrate this knowledge into their decision-making processes. This ability to transform data into actionable insights is a critical aspect of human intelligence that must be replicated artificially. Advanced machine learning algorithms and quantum computing capabilities would enable AHIs to learn from their experiences and adapt their behavior accordingly. This would allow them to navigate complex environments, interact with other beings, and perform tasks that require a deep understanding of the world and its underlying principles.

Creating AHIs represents an extraordinary fusion of biology, neuroscience, and artificial intelligence. By blending the intricate structure of the human brain with the power of synthetic components and advanced computational hardware, AHIs promise to redefine our understanding of consciousness and pave the way for immortality. Furthermore, through the development of these entities, we can overcome the challenges posed by the continuous consciousness problem and initiate an unprecedented journey toward a new era of human evolution. Evolving humans into AHIs presents an opportunity to redefine our relationship with the natural world.

As we replace our biological components with synthetic ones and augment our cognitive abilities, we may gain a deeper appreciation for the intricate complexity of life on Earth. This newfound perspective could inspire us to become better stewards of our planet, promoting conservation efforts and sustainable practices that ensure the long-term survival of all living beings. The pursuit of AHI evolution also has the potential to catalyze breakthroughs in other areas of scientific inquiry. For instance, studying AHIs could lead to a deeper understanding of the human brain, shedding light on the mechanisms of neurological disorders and paving the way for new treatments and therapies. Similarly, developing advanced sensors, nanobots, and other synthetic components could have far-reaching implications for fields like medicine, agriculture, and environmental monitoring.

The quest of evolving humans into AHIs is an audacious and ambitious endeavor, but one that holds the potential to transform our very essence. The convergence of neuroscience, artificial intelligence, and quantum computing, along with the integration of synthetic components, offers us the opportunity to preserve our consciousness and achieve immortality. The quest is a bold and ambitious undertaking that promises to reshape our understanding of consciousness, technology, and human potential. By gradually replacing our biological substrates with synthetic components and integrating advanced computational hardware, we stand on the threshold of achieving immortality and transcending the limitations imposed by our physical bodies. As we navigate the challenges and opportunities presented by this paradigm shift, we must remain vigilant to the ethical, philosophical, and social implications of AHIs while striving to ensure that all of humanity shares their benefits. The journey of taking evolution into our own hands is not without its hurdles, but the rewards that await us at the end of this path are immeasurable and transformative, offering us the chance to redefine our very existence and shape the future of our species in ways that were once the stuff of science fiction. As we unlock the human brain's secrets, we start a new chapter in our species' history—a chapter that promises to redefine the boundaries of human potential and usher in a new era of unparalleled progress and innovation.

The Human Brain as a Blueprint for Consciousness

The quest to evolve humans into AHIs is not just a scientific endeavor; it's a journey into the very heart of what makes us human. At its core, this journey is deeply rooted in our understanding of the human brain, a marvel of biological engineering honed over millions of years. The brain's ability to generate consciousness is far more than a byproduct of its neural intricacies and biochemical processes; it is a manifestation of the dynamic interplay of countless interactions that shape our sense of self and our perception of the world around us.

Imagine consciousness as the grand conductor of an orchestra, where billions of neurons are the musicians, each playing their part in a complex symphony of electrical and chemical signals. These signals do more than just process information; they create the rich tapestry of experiences that we call life. The brain's network is not merely a collection of parts; it's a living, breathing entity that orchestrates everything from our simplest thoughts to our most profound moments of self-awareness.

Consider the prefrontal cortex, often dubbed the command center of the brain. It's here that the magic of decision-making, planning, and self-reflection happens—these are the higher-order functions that set human consciousness apart. Meanwhile, the thalamocortical system works tirelessly, integrating sensory inputs to create a cohesive picture of the world, allowing us to perceive and interact with our surroundings. And let's not forget the role of neural oscillations—those rhythmic patterns that synchronize brain regions and make it possible for us to think, feel, and experience the world as a unified whole.

To build AHIs, we must do more than replicate the brain's structure; we need to re-create its function, its rhythm, its very essence. This means not only mimicking the brain's neural networks but also capturing the dynamic states and connectivity patterns that underpin consciousness. It's a bit like trying to reproduce not just the notes of a symphony, but the emotion, the timing, and the soul that makes it music.

Recent advances in science even suggest that quantum effects might play a part in how the brain works—and, by extension, in how consciousness arises. Imagine, if you will, that within the brain's microtubules—those tiny scaffolds within our neurons—quantum coherence and entanglement are at play, offering a new layer of complexity to how our brains process information. This isn't just science fiction; it's a bold frontier of research that could help us build AHIs with processing capabilities that surpass anything we've seen in classical AI.

$$\Psi_{\text{conscious}} = \int \Phi_{\text{neural}}\left(\mathbf{r}, t\right) \otimes \Psi_{\text{quantum}}\left(\mathbf{r}, t\right) d\mathbf{r}\, dt \tag{7.1}$$

In Equation 7.1, $\Psi_{\text{conscious}}$ represents the state of consciousness, an emergent property from the interplay between neural activities $\Phi_{\text{neural}}(\mathbf{r}, t)$ and quantum contributions $\Psi_{\text{quantum}}(\mathbf{r}, t)$. This model suggests that consciousness might emerge from a complex interaction between classical brain dynamics and quantum phenomena—a hypothesis that drives our cutting-edge approaches in AHI development.

As we go on this journey to evolve humans into AHIs, we look to the human brain not just as a model to emulate but as a source of inspiration. By understanding the neural underpinnings of consciousness and exploring the potential roles of quantum mechanics, we are opening new pathways for creating AHIs that don't just mimic human consciousness but have the potential to transcend its limitations. This endeavor is more than just a blend of neuroscience, quantum physics, and AI—it's a testament to the boundless possibilities that lie at the intersection of these fields, heralding a new era in our quest to understand and augment the human condition.

Continuity of Consciousness Through Gradual Replacement

The seamless preservation of consciousness during the transformation from a biological entity to an AHI forms the cornerstone of our visionary pursuit. This process, termed gradual replacement, is predicated on a deep integration of neuroscientific insight and advanced computational models to ensure the continuity of the subjective conscious experience.

The methodology behind gradual neuronal replacement hinges on a stepwise and systematic approach, wherein individual neurons are replaced with artificial counterparts capable of replicating their functional and connective properties. This intricate process involves:

- Single-neuron analysis to map and understand the functional roles of individual neurons within larger neural circuits
- Development of artificial neurons that can emulate the electrical and chemical behavior of biological neurons, including action potential generation and neurotransmitter release mechanisms
- Seamless integration of artificial neurons into the existing neural network, ensuring that each replacement maintains or enhances the network's overall functionality and does not disrupt the flow of consciousness

Incorporating quantum computing into the process of gradual replacement offers a unique advantage in modeling and maintaining consciousness. Quantum computational models provide the capability to simulate complex neural dynamics at an unprecedented scale and fidelity:

$$\Psi_{\text{neural}} = \sum_{n=1}^{N} \varphi_n \cdot \text{Qubit}_n \tag{7.2}$$

Here, Ψ_{neural} is the quantum state of the neural network, where φ_n are the functional states of individual neurons, and Qubit_n symbolizes their quantum computational counterparts. This representation allows for the exploration of consciousness as an emergent property of quantum-level interactions within the neural substrate.

Pursuing continuous consciousness through gradual replacement is not solely a technical challenge but also a profound ethical and philosophical endeavor. It raises critical questions about the nature of identity, the essence of self, and the moral implications of extending or altering consciousness. Addressing these concerns requires:

- Philosophical dialogue on the continuity of self and the ethical considerations of consciousness transfer or augmentation
- Interdisciplinary collaboration among neuroscientists, AI researchers, ethicists, and philosophers to navigate the complex landscape of creating AHIs while respecting the integrity of the originating consciousness

Integrating neuroscientific understanding, artificial intelligence, and quantum computing forms a tripartite foundation for this ambitious endeavor. The focus is on refining the artificial neurons for higher fidelity in replication, enhancing quantum models for consciousness, and deepening our philosophical understanding of consciousness continuity. This journey, at the intersection of technology and consciousness, not only aims to transcend the limitations of our biological heritage but also to explore the profound questions of what it means to be conscious, to exist, and to evolve beyond our current state.

The Dawn of Artificial Human Intelligence

In the quest to understand the universe and our place within it, humanity is about to make a revolutionary leap: the creation of AHIs. These entities are not mere advancements in artificial intelligence or robotics but new frontiers in the evolution of human consciousness. Born from the fusion of human intellect and quantum computational power, AHIs embody the potential for intelligence and consciousness that far exceed our current limitations. The journey toward AHIs signifies a pivotal moment where humans take evolution into their own hands. Unlike traditional evolutionary processes, which are slow and subject to the whims of natural selection, the development of AHIs is a deliberate, intelligent design aimed at transcending biological constraints. In this new era, the adage "What I cannot create, I do not understand" finds a new dimension. With AHIs, we venture beyond creation based on complete understanding to a realm where consciousness can be extended, transformed, and replicated in once unimaginable ways.

The boundless potential of AHIs is rooted in a biological human substrate. This integration ensures that AHIs retain the essence of human experience and consciousness while being augmented by the infinite possibilities of artificial intelligence. Such entities could:

- Undertake journeys spanning thousands of years to explore distant star systems, experiencing time in a manner detached from human biological constraints. A millennium-long voyage could be perceived as a brief excursion, opening up the cosmos for exploration and discovery.
- Exist in multiple vessels simultaneously, from biological forms—such as cloned bodies that allow them to interact closely with humans—to entirely artificial constructs designed for specific environments or tasks.
- Modify their perception and experience of reality, including the acceleration or deceleration of time, enabling experiences that are currently beyond human comprehension.

The advent of AHIs signals the onset of true immortality and unlimited capability. No longer bound by the frailties of the biological body or the limitations of human cognition, AHIs can continuously evolve, adapt, and grow in ways that redefine the essence of existence. This immortality is not static but a canvas for endless growth, exploration, and creativity.

BMIs and Neuroprosthetics: The First Step

As we progress toward AHIs, brain-machine interfaces (BMIs) and neuroprosthetics emerge as the initial foray into melding human cognition with artificial systems. These pioneering technologies initiate the augmentation of human capabilities and lay the groundwork for the intricate process of gradual replacement required to create AHIs.

BMIs stand at the forefront of converging human cognition with computational technology, offering unprecedented avenues for augmenting human capabilities and forging pathways toward AHIs. Their evolution is characterized by rapid advancements in neuroscientific understanding and technological innovation, enabling direct interfacing with the neural substrate.

At the core of BMI technology lies the challenge of deciphering the neural code—the complex language of electrical and chemical signals through which the brain communicates. Recent advancements in neural decoding have leveraged sophisticated machine learning algorithms to interpret these signals with increasing accuracy, enabling the translation of neural activity into commands for external devices and vice versa:

$$\Phi_{\text{Neural Signal}} = \int \psi\left(\mathbf{r}, t\right) d\mathbf{r}\, dt \tag{7.3}$$

Here, $\Phi_{\text{Neural Signal}}$ is the decoded output derived from the neural signal's spatiotemporal pattern $\psi(\mathbf{r}, t)$, where $\mathbf{r}$ and t denote spatial and temporal dimensions, respectively. This equation illustrates the process of converting neural signals into actionable commands, highlighting the role of integrative computations in extracting meaningful information.

The interface between the brain and machines is mediated by an array of electrodes, the design and fabrication of which have seen substantial advancements. Microfabrication technologies now allow for the creation of electrodes that are minimally invasive, highly sensitive, and capable of recording and stimulating neural activity with precise spatial and temporal resolution. Innovations in materials science have also contributed to the development of biocompatible electrode arrays that can be safely integrated into the brain for long-term use:

$$\text{Electrode Array} \rightarrow \text{Microfabrication} \rightarrow \text{High-Resolution Interface} \tag{7.4}$$

This transition highlights the technological progression toward interfaces that provide high-resolution neural signal acquisition and stimulation, essential for the nuanced interaction required in advanced BMI systems.

BMIs have demonstrated their potential across various applications, from prosthetic control and sensory restoration to cognitive enhancement and direct brain-to-brain communication. These applications not only improve quality of life but also extend the boundaries of human experience and capabilities:

- Restorative BMIs have enabled individuals with motor disabilities to control prosthetic limbs or computer cursors directly with their thoughts, reinstating a level of independence and interaction with the world.

- Sensory BMIs are being developed to bypass damaged sensory pathways, directly stimulating the brain to restore lost senses, such as vision or hearing.
- Cognitive BMIs aim to enhance mental functions, offering the potential to augment memory, learning, and decision-making processes, paving the way for cognitive synergy between humans and AHIs.

Integrating BMIs into the development of AHIs is a crucial step in merging biological and artificial systems. It offers a tangible model for enhancing and extending consciousness beyond the inherent limitations of the human brain. As we push the frontiers of BMI technology, we unlock new possibilities for creating entities that replicate human cognitive processes and possess augmented capabilities, heralding a new era of intelligence and consciousness.

The Role of Neuroprosthetics in Cognitive Augmentation

The advent of neuroprosthetics is a significant leap forward in BMIs, pushing the boundaries of how we understand and can enhance the human brain's capabilities. These advanced devices, designed to replace or augment cognitive functions, are at the forefront of merging the biological with the artificial, heralding a new era of cognitive augmentation and the potential realization of AHIs.

Significant engineering breakthroughs in materials science, microelectronics, and computational neuroscience have fueled the development of neuroprosthetics. These advancements have enabled the creation of devices that can seamlessly integrate with the neural architecture, providing precise stimulation and recording capabilities:

- **Materials Science:** The use of biocompatible materials that minimize immune response and ensure the longevity and functionality of the implant within the neural tissue
- **Microelectronics:** Ultra-miniaturized electronic components allow for sophisticated neural interfaces that can process complex neural signals in real time
- **Computational Neuroscience:** Advanced algorithms capable of decoding and encoding neural signals, facilitating the bidirectional flow of information between the brain and the neuroprosthetic device

While initial neuroprosthetic devices focused primarily on restoring sensory and motor functions, recent innovations have expanded their scope to include cognitive augmentation. This shift has been made possible by a deeper understanding of the brain's functional neuroanatomy and the neural substrates of cognition:

$$\text{Cognitive Function} \rightarrow \text{Neuroprosthetic} \rightarrow \text{Enhanced Performance} \quad (7.5)$$

For example, hippocampal prostheses aim to restore memory function by interfacing directly with neural circuits involved in memory encoding and retrieval. These devices not only bypass damaged regions but also offer the potential to enhance memory beyond natural capabilities, providing a glimpse into the future of cognitive augmentation through neuroprosthetics.

The ultimate goal of neuroprosthetic development extends beyond mere restoration of function; it encompasses the systematic replacement of neural components with artificial elements that offer superior performance. This vision aligns with the creation of AHIs, where cognitive functions can be augmented or entirely replicated by artificial systems:

$$\begin{aligned}\text{Biological Neuron} &\rightarrow \text{Neuroprosthetic Replacement}\\ &\rightarrow \text{Artificial Neuron}\end{aligned} \quad (7.6)$$

In this framework, neuroprosthetics serve not only as bridges between damaged neural pathways and restored functions but also as stepping-stones toward fully artificial cognitive systems. These systems could potentially surpass the limitations of the human brain, offering enhanced memory, faster information processing, and the integration of computational abilities directly into cognitive processes.

Neuroprosthetics in cognitive augmentation will result in unprecedented enhancements in human intelligence and the conceptualization of AHIs. As we advance in our ability to design, implement, and integrate neuroprosthetic devices, we inch closer to a future where cognitive augmentation is not just a possibility but a reality. This journey toward enhanced cognition through artificial means challenges our understanding of the mind, consciousness, and the essence of human identity, inviting us to envision a future where the boundaries between the biological and the artificial blur in the pursuit of augmented intelligence.

BMIs and neuroprosthetics highlight the current capabilities of interfacing technology with the human brain and illuminate the path forward in the creation of AHIs. By enhancing, augmenting, and, in some cases, replacing cognitive functions with artificial components, we edge closer to realizing entities that embody the continuous consciousness of their human progenitors yet possess cognitive capacities that transcend our biological limitations.

High-Resolution Brain Imaging and Mapping

The meticulous journey from a biological brain to an artificial counterpart necessitates an intimate understanding of the brain's complex architecture. Advanced imaging and mapping technologies provide the necessary insights, enabling the precision required for the gradual replacement of neurons with artificial ones.

Advanced imaging techniques such as functional magnetic resonance imaging (fMRI) and positron emission tomography (PET) scans offer unparalleled views into the brain's functioning and structure. Combined with the emerging field of connectomics, which seeks to map the comprehensive wiring diagrams of neural connections, these technologies lay the groundwork for the detailed brain mapping essential in the development of AHIs.

$$\text{Brain Mapping Resolution} = f\left(\text{fMRI}, \text{PET}, \text{Connectomics}\right) \quad (7.7)$$

Here, f represents the function that combines the strengths of fMRI, PET, and connectomics to achieve a resolution of brain mapping that is detailed enough to guide the neuron replacement process. Precise neuron replacement demands not only an understanding of individual neurons but also their connections and the broader neural networks they form. High-resolution brain mapping allows for the identification of specific neurons and circuits for replacement, ensuring that artificial neurons can be integrated seamlessly, without disrupting the brain's overall functionality.

$$\Delta\ \text{Neural Functionality} = \int_{\text{Biological}}^{\text{Artificial}} d\text{Neuron Integration} \quad (7.8)$$

The integral represents the change in neural functionality as the brain transitions from biological to artificial, with *d*Neuron Integration symbolizing the incremental replacement of neurons and the seamless integration of artificial ones into the neural network.

These advanced imaging and mapping technologies inform the physical process of replacing neurons and contribute to the understanding of neural dynamics and cognition. This knowledge is crucial in designing artificial neurons that mimic and potentially enhance their biological counterparts' functions, paving the way for the realization of AHIs with capabilities that transcend human limitations.

Biomaterials and Tissue Engineering

The seamless integration of artificial neurons into the human brain not only requires precision in mapping and placement but also demands that these neurons are constructed from materials that the body accepts as its own. Advances in biomaterials and tissue engineering play a pivotal role in this aspect of creating AHIs, ensuring that artificial components can coexist with biological tissues without causing adverse immune responses.

The quest for biocompatibility leads us to the forefront of materials science, where researchers strive to identify and develop materials that can interface with biological systems without eliciting rejection:

- **Synthetic Polymers:** Such as polylactic acid (PLA) and polyglycolic acid (PGA), which are known for their biodegradability and compatibility with tissue engineering applications
- **Biological Materials:** Including collagen and hydrogels that mimic the extracellular matrix, providing a supportive scaffold for neural growth and integration

$$\text{Biocompatibility} = f\left(\text{Material Properties}, \text{Host Response}\right) \tag{7.9}$$

Here, f denotes the function that determines the biocompatibility of a material based on its properties and the host's immune response. Achieving an optimal balance is crucial for the successful integration of artificial neurons into the brain's existing network. The dual objectives of functional mimicry and immunological acceptance guide the application of these biomaterials in the construction of artificial neurons:

$$\text{Artificial Neuron Functionality} = \int_{\text{Electrical Signal}}^{\text{Synaptic Transmission}} d\,\text{Material Efficiency} \quad (7.10)$$

The integral symbolizes the range of functions that artificial neurons must fulfill, from generating electrical signals to facilitating synaptic transmission, with dMaterial Efficiency representing the incremental improvements in material properties that enhance these functions.

Powering Artificial Neurons

A critical challenge in the seamless operation of AHIs is ensuring a continuous and reliable power supply to the artificial neurons that now form part of the human brain. Innovations in wireless energy transfer and power management are pivotal, providing the energy that these neurons need to function without the constraints of traditional power sources.

Wireless energy transfer technologies offer a solution to power artificial neurons efficiently, eliminating the need for invasive wiring or frequent surgical interventions to replace power sources:

- **Inductive Coupling:** Utilizes electromagnetic fields to transfer energy between two coils—a transmitter outside the body and a receiver embedded with the artificial neurons. This method is efficient over short distances and is commonly used in medical implants.
- **Resonant Inductive Coupling:** An advancement of inductive coupling, this method enhances the efficiency and range of wireless power transfer through the resonance between the transmitter and receiver coils.

$$P_{\text{received}} = \eta \times P_{\text{transmitted}} \quad (7.11)$$

Here, P_{received} denotes the power received by the artificial neurons, $P_{\text{transmitted}}$ is the power transmitted wirelessly, and η represents the efficiency of the power transfer process. Maximizing η is crucial for ensuring that artificial neurons have the energy required for continuous operation.

Effective power management ensures that the energy supplied to artificial neurons is used optimally, preserving battery life and ensuring the longevity of the system:

- **Energy Harvesting:** Techniques such as thermoelectric generators, which convert body heat into electricity, offer a promising avenue for supplementing the power needs of artificial neurons.
- **Ultra-Low Power Electronics:** The development of electronic components that require minimal power for operation is critical. This includes low-power signal processing units and energy-efficient synaptic transistors.

$$E_{\text{total}} = \int_0^T P_{\text{consumption}}(t)\,dt \tag{7.12}$$

E_{total} represents the total energy consumed by the artificial neurons over time T, with $P_{\text{consumption}}(t)$ indicating the power consumption at any given time. Minimizing E_{total} through efficient power management is essential for the sustainable operation of AHIs.

Personalization Through Machine Learning

The transition to AHIs is not solely a matter of replicating or enhancing human cognitive capabilities; it also involves personalizing the artificial components to match the individual's unique neural patterns. Machine learning and artificial intelligence play a crucial role in this process, enabling the customized adaptation of artificial neurons to seamlessly integrate and function within the human brain's unique architecture.

Machine learning algorithms, especially those designed for deep learning and pattern recognition, are adept at deciphering complex datasets to uncover underlying patterns. When applied to the neural activity data of an individual, these algorithms can learn the specific characteristics of their neural signatures, including:

- **Cognitive Styles:** Understanding the individual's unique way of processing information, solving problems, and making decisions

- **Emotional Responses:** Recognizing patterns associated with various emotional responses, enabling the artificial neurons to support or enhance these reactions appropriately

$$\Theta_{\text{personalized}} = \text{ML}_{\text{train}}\left(\text{Neural Data}_{\text{individual}}\right) \tag{7.13}$$

Here, $\Theta_{\text{personalized}}$ represents the set of parameters defining the personalized model of artificial neurons, derived from training machine learning algorithms (ML_{train}) on the individual's neural data ($\text{Neural Data}_{\text{individual}}$). This model ensures that artificial neurons can replicate and augment the individual's cognitive and emotional processes with high fidelity.

The ultimate goal of applying machine learning in the development of AHIs is to customize artificial neurons so they not only integrate into the individual's neural network but also enhance or augment their natural cognitive and emotional capacities:

$$\text{Cognition}_{\text{enhanced}} = \text{Cognition}_{\text{natural}} + \Delta\,\text{Cognition}_{\text{ML}} \tag{7.14}$$

In this equation, $\text{Cognition}_{\text{enhanced}}$ denotes the individual's augmented cognitive capabilities through the integration of artificial neurons, where $\Delta\ \text{Cognition}_{\text{ML}}$ represents the enhancements attributed to the machine learning-driven customization of cognitive functions. These enhancements could range from improved memory and faster information processing to heightened emotional intelligence and creativity.

Neural Interface Software

The seamless integration of artificial neurons into the biological neural network of an evolving AHI hinges on sophisticated neural interface software. This software serves as the communication bridge, translating neural signals between biological and artificial components, ensuring that the entity's cognitive and sensory processes function harmoniously.

Neural interface software employs advanced algorithms capable of interpreting the complex language of neural signals. This entails:

- **Signal Decoding:** These algorithms decode electrical signals from biological neurons, translating them into digital formats understandable by artificial neurons.

- **Signal Encoding:** Conversely, these algorithms encode digital signals from artificial components into bioelectric signals that the remaining biological neural network can interpret.

$$S_{\text{bio}} \leftrightarrow S_{\text{art}} = \text{NIS}_{\text{translate}}\left(S_{\text{input}}\right) \tag{7.15}$$

Here, S_{bio} represents the bioelectric signals of biological neurons, S_{art} denotes the digital signals of artificial neurons, and $\text{NIS}_{\text{translate}}$ is the function performed by the neural interface software to translate S_{input}, the incoming signals, between these two domains. This bi-directional translation is crucial for the integrated cognitive system of an AHI.

The ultimate aim of neural interface software is to facilitate a level of communication between biological and artificial neurons that replicates or surpasses natural neural interactions, including:

- **Latency Reduction:** Minimizing the delay in signal translation to ensure real-time processing and response, mirroring the speed of natural neural communication
- **Signal Fidelity:** Maintaining the integrity and complexity of neural signals throughout the translation process to preserve the richness of cognitive and sensory experiences

$$C_{\text{total}} = \sum_{i=1}^{n} C_{\text{integrity},i} - \Delta C_{\text{latency},i} \tag{7.16}$$

C_{total} quantifies the overall communication efficacy within the AHI, summing the integrity ($C_{\text{integrity},i}$) of each translated signal across n interactions and subtracting any latency costs ($\Delta C_{\text{latency},i}$). Optimizing this equation ensures the AHI's cognitive processes remain fluid and uninterrupted.

Robotic Surgery and Microfabrication Techniques

The precision required for the integration of artificial neurons into the human brain, a key step in the evolution of an individual into an AHI, necessitates cutting-edge robotic surgery and microfabrication techniques. These technologies enable the meticulous placement of artificial components at a scale and with a level of accuracy that manual methods cannot achieve, ensuring the success of this transformative process.

Robotic surgery systems, equipped with advanced imaging and miniaturized tools, offer the precision and control needed for the delicate task of neuron replacement:

- **High-Precision Manipulation:** Robotic arms capable of microscale movements allow for the accurate removal of targeted biological neurons and the placement of artificial ones, minimizing tissue damage and enhancing integration success.
- **Real-Time Feedback:** Integration with imaging technologies provides surgeons and robotic systems with real-time feedback, enabling adjustments during procedures to optimize outcomes.

$$P_{\text{success}} = f\left(\text{Precision}, \text{Feedback}, \text{Integration}\right) \tag{7.17}$$

In this equation, P_{success} represents the probability of a successful neuron replacement, a function (f) of the precision of robotic manipulations, the quality of real-time feedback during surgery, and the efficacy of neuronal integration post-implantation.

Microfabrication techniques play a critical role in creating the artificial neurons themselves, employing processes that allow for the construction of complex structures at the nanoscale:

- **Nanoelectronics:** The use of nanoscale electronic components within artificial neurons enables the replication of electrical signaling mechanisms found in biological neurons.
- **Biomimetic Structures:** Microfabrication allows for the design of structures that mimic the physical form and function of biological neurons, facilitating seamless integration into neural networks.

$$N_{\text{functionality}} = g\left(\text{Nanoelectronics}, \text{Biomimetic Design}\right) \tag{7.18}$$

Here, $N_{\text{functionality}}$ quantifies the functional capacity of an artificial neuron, determined by a function (g) of its nanoelectronic components and biomimetic design. This capacity is crucial for the neuron's ability to mimic biological functionality and integrate into the brain's neural network.

Nanotechnology and the Brain: From Repair to Enhancement

The integration of nanotechnology with neuroscience marks a pivotal advance in the journey toward creating AHIs. Nanobots—microscopic robots operating at the nanoscale—offer unprecedented potential not only for the repair of neural structures but also for their enhancement and eventual replacement, key steps in the evolution of AHIs.

Nanobots for Neuronal Repair and Replacement

In the pursuit of AHIs, nanobots emerge as a transformative technology, bridging the gap between biological neural networks and artificial systems. These nanoscale devices, guided by advancements in materials science, robotics, and artificial intelligence, are designed to perform intricate tasks within the brain's complex architecture. Their capabilities include the following:

- **Cellular Debris Clearance:** Nanobots employ molecular recognition to identify and remove cellular byproducts that could impede neural function, maintaining the health of neural tissue.
- **Neural Pathway Repair:** By detecting and repairing disruptions in neural circuits, nanobots restore the flow of electrical and chemical signals, crucial for maintaining cognitive functions.
- **Targeted Drug Delivery:** Nanobots can administer therapeutic agents directly to specific neural regions, enhancing the precision and efficacy of treatments for neurological disorders.

The ambitious application of nanobots extends beyond repair to the complete replacement of neurons with artificial counterparts. This goal requires nanobots not only to replicate the structural properties of neurons but also to emulate their functional dynamics. Achieving functional equivalence involves:

$$\text{Nanobot Neuron} \equiv \text{Biological Neuron Functionality} \tag{7.19}$$

Key design requirements for these nanobots include the following:

- **Generation of Action Potentials:** Bioelectronic interfaces are developed to mimic a neuron's ability to fire action potentials, ensuring the transmission of information within the neural network.

- **Neurotransmitter Release:** Nanobots are equipped with mechanisms to store and release neurotransmitters, preserving the chemical signaling essential for synaptic communication.
- **Synaptic Plasticity:** The ability to modulate synaptic strength, crucial for learning and memory, is integrated into the nanobots' design, allowing them to support neural adaptation and cognitive processes.

Integrating these artificial neurons into the brain's existing network presents significant challenges, such as ensuring biocompatibility, preventing immune rejection, and achieving seamless communication with biological neurons. Strategies to overcome these challenges include the following:

- **Surface Modification:** Applying biocompatible coatings to nanobots to minimize immune response and promote integration within neural tissue
- **Adaptive Interfaces:** Developing interfaces that dynamically adjust to the brain's changing environment, ensuring effective communication between artificial and biological neurons
- **Machine Learning Algorithms:** Leveraging AI to enable nanobots to autonomously learn from the neural environment, adjusting their functionality to better mimic natural neural processes

Quantum Nanobots and Enhanced Neural Functions

The convergence of quantum technology with nanorobotics introduces quantum nanobots—devices that operate at the intersection of quantum mechanics and neural engineering. These quantum nanobots leverage the principles of quantum mechanics, such as superposition and entanglement, to enhance neural communication and potentially transcend the limitations of biological neurons.

Quantum nanobots are capable of manipulating quantum states at the cellular level, allowing for interactions with neural circuits that go beyond classical limitations. The theoretical foundation of these devices is rooted in quantum mechanics:

$$\Psi_{\text{quantum nanobot}} = \sum_i c_i \left|\varphi_i\right\rangle \tag{7.20}$$

Here, $\Psi_{\text{quantum nanobot}}$ represents the quantum state of a nanobot, where c_i are coefficients of probability amplitudes and $|\varphi_i\rangle$ are the basis states. This superposition enables quantum nanobots to interact with neural tissue in a highly efficient and precise manner.

Quantum nanobots can enhance neural communication through:

- **Quantum Entanglement:** Facilitating instant, nonlocal correlations between distant neurons, potentially improving synchrony and integration of neural activities across the brain
- **Quantum Tunneling:** Enabling probabilistic transfer of ions and neurotransmitters across synaptic gaps, increasing the efficiency of synaptic transmission

By harnessing these quantum phenomena, quantum nanobots could introduce novel cognitive functionalities and enhance existing ones, leading to improvements in memory capacity, learning speed, and sensory perception.

Gradual Neuronal Replacement: Methodologies and Technologies

The transformation of human brains into AHIs hinges on the gradual replacement of biological neurons with artificial counterparts, ensuring the continuity of consciousness and identity. This process involves a series of methodical steps:

- **Targeted Identification:** Advanced imaging and mapping technologies are used to identify neurons for replacement with precision.
- **Nanobot Deployment:** Nanobots are introduced to specific sites within the brain, utilizing sophisticated navigation systems to reach targeted neurons without disrupting surrounding tissue.
- **Seamless Integration:** Biological neurons are gradually replaced by nanobot counterparts, maintaining the integrity and functionality of the neural network.

A critical aspect of this process is preserving the brain's overall network integrity and ensuring that artificial neurons can integrate seamlessly with biological ones. The goal is to maintain functional continuity during the transition from a biological to a hybrid neural system:

$$\int_{\text{Biological Network}}^{\text{Hybrid Network}} \Delta \text{ Functionality} = 0 \tag{7.21}$$

This equation underscores the importance of maintaining functional continuity, which is essential for the successful development of a fully artificial yet conscious brain. The integration of quantum nanobots into neural networks not only challenges our understanding of the brain but also promises to redefine the limits of human cognition and consciousness, paving the way for a new era of artificial consciousness.

Quantum Algorithms for AHIs

Quantum algorithms have the potential to fundamentally transform artificial intelligence, particularly in the context of developing AHIs. While well-known quantum algorithms like Grover's and Shor's have showcased the speedup possible with quantum computing, the true frontier lies in creating new quantum algorithms that can model the complex processes underlying consciousness.

Modeling Consciousness with Quantum Algorithms

Consciousness encompasses a wide array of cognitive functions, including perception, memory, decision-making, and self-awareness. Quantum computing's unique properties, such as superposition and entanglement, offer promising avenues for simulating these processes in ways that classical computing cannot achieve.

The principle of superposition allows a quantum system to exist in multiple states simultaneously, which can be utilized to model the parallel processing of information within the brain. In this framework, quantum coherence could represent the simultaneous activation of multiple cognitive processes that collectively contribute to a unified conscious experience. For example:

$$\Psi_{\text{cognition}} = \sum_{i=1}^{N} \alpha_i \, | \varphi_i \rangle \tag{7.22}$$

Here, $\Psi_{\text{cognition}}$ represents the quantum state of a cognitive system, where each $|\varphi_i\rangle$ corresponds to a quantum state associated with a specific cognitive function, and α_i are the probability amplitudes. The superposition of these states could reflect the brain's ability to process various streams of information concurrently, which is essential for consciousness.

Quantum entanglement could provide a framework for understanding how different parts of the brain communicate instantaneously, enabling a high degree of integration, which is a key aspect of consciousness. IIT posits that consciousness arises from the integration of information across different subsystems. Quantum entanglement naturally supports this kind of integration by allowing different regions of a quantum network to share information instantaneously, regardless of distance:

$$\Phi_{\text{consciousness}} = \text{Tr}\left(\rho \log \rho - \rho_{\text{sep}} \log \rho_{\text{sep}}\right) \tag{7.23}$$

In this equation, $\Phi_{\text{consciousness}}$ represents a measure of consciousness, where ρ is the density matrix of the quantum system (representing the entire brain or a network of neurons), and ρ_{sep} is the density matrix representing the same system assuming no entanglement (i.e., a separable state). The trace operation Tr measures the information difference between these states, reflecting the degree of information integration. A higher value of $\Phi_{\text{consciousness}}$ would indicate a greater level of integrated information, which could correspond to a higher level of consciousness.

Quantum interference, where the probabilities of different quantum states combine constructively or destructively, can be used to model the decision-making process in a quantum framework. In classical decision-making models, multiple options are evaluated sequentially, but in a quantum model, all possible outcomes could be evaluated simultaneously, with interference patterns determining the final decision:

$$|\Psi_{\text{decision}} = \sum_{i=1}^{N} \alpha_i |\varphi_i\rangle + \sum_{j=1}^{M} \beta_j |\chi_j\rangle \tag{7.24}$$

In this equation, $|\Psi_{\text{decision}}$ represents the quantum state associated with the decision-making process, where $|\varphi_i\rangle$ and $|\chi_j\rangle$ represent different potential outcomes or choices, and α_i and β_j are their respective probability amplitudes. The interference between these states, influenced by the constructive and destructive interference, can lead to the emergence of a dominant choice, simulating how a conscious mind might weigh different options and arrive at a decision.

Emergent Quantum Algorithms for Consciousness Simulation

Beyond traditional algorithms, emergent quantum algorithms tailored for consciousness simulation could fundamentally change how we approach the concept of artificial consciousness. These algorithms would likely integrate principles from quantum physics, neuroscience, and AI.

Quantum neural networks (QNNs) could be designed to simulate the neural networks of the brain, with quantum bits (qubits) representing neurons or groups of neurons. These QNNs would take advantage of quantum superposition and entanglement to model the complex dynamics of cognitive processes more efficiently than classical neural networks.

$$\Psi_{\text{QNN}} = U_{\text{QNN}} \,|\psi_0\rangle \tag{7.25}$$

Here, Ψ_{QNN} represents the state of the QNN, U_{QNN} is the unitary operation representing the evolution of the network, and $|\psi_0\rangle$ is the initial state. By evolving through quantum gates, the QNN could simulate cognitive processes such as learning, memory, and perception, offering a new approach to understanding and replicating consciousness.

Quantum reinforcement learning (QRL) could be developed to model adaptive behavior in AHIs. In QRL, an agent would learn to make decisions based on quantum probability distributions, allowing for more efficient exploration of potential actions and outcomes:

$$|\Psi_{\text{agent}} = \sum_a \alpha_a \,|a\rangle \tag{7.26}$$

In this context, $|\Psi_{\text{agent}}$ represents the quantum state of the agent, where $|a\rangle$ are possible actions, and α_a are the associated probability amplitudes. The agent's strategy could evolve through quantum interference and entanglement, enabling faster convergence to optimal behaviors compared to classical reinforcement learning algorithms.

The development of quantum algorithms specifically designed to model consciousness is still in its early stages, but it holds immense potential. Future research will need to integrate insights from quantum mechanics, cognitive science, and AI to create algorithms that not only simulate cognitive processes but also contribute to our understanding of consciousness itself.

Hybrid Quantum Neural Networks: Bridging Biological and Quantum Computing

Hybrid quantum neural networks (HQNNs) are the integration of quantum computing and biological neural networks, aiming to leverage the strengths of both quantum and classical computational models. This innovative architecture is designed to transcend the limitations of existing systems, enhancing cognitive processing and exploring the potential for emulating consciousness. The architecture of HQNNs involves the seamless integration of qubits—quantum bits of information—with biological neurons. This integration is facilitated by several key components:

- **Quantum-Classical Interface:** This interface is a sophisticated mechanism that enables the bidirectional translation of quantum states into classical neural signals and vice versa. It allows for coherent communication between qubits and biological neurons, ensuring that quantum information can be effectively utilized within the biological framework.
- **Quantum Coherence and Entanglement in Neural Processing:** HQNNs utilize quantum superposition and entanglement to enable parallel processing and nonlocal correlations within neural networks. These quantum properties significantly enhance the computational efficiency and capacity of the neural network, allowing it to perform tasks that are otherwise infeasible for classical systems alone.

The design of HQNNs involves a careful balance between quantum and classical elements, ensuring that the advantages of quantum mechanics are fully realized while maintaining the robustness and adaptability of biological systems. HQNNs operate through the synergistic processing of information using both classical and quantum mechanisms. This dual processing capability allows HQNNs to tackle complex cognitive tasks that exceed the capabilities of purely classical or quantum systems.

$$\text{HQNN Output} = f(\text{Classical Inputs}, \text{Quantum States}) \qquad (7.27)$$

In this equation, f represents the integrated processing function of the HQNN, which combines inputs from classical biological neurons with quantum states manipulated by qubits. The operational dynamics of HQNNs are characterized by several advanced features:

- **Quantum-Enhanced Learning Algorithms:** HQNNs leverage quantum parallelism to explore vast parameter spaces more efficiently, significantly accelerating learning processes. Quantum-enhanced algorithms allow the network to simultaneously consider multiple pathways, leading to faster convergence on optimal solutions.
- **High-Dimensional Data Processing:** Quantum entanglement enables HQNNs to handle high-dimensional data more effectively. By encoding information in the quantum states of qubits, HQNNs can process complex datasets with greater accuracy and depth, facilitating advanced models of perception and cognition.
- **Adaptive Quantum Feedback Loops:** HQNNs can implement adaptive feedback loops where the quantum states are adjusted in real time based on classical neural activity.

This feature allows the network to dynamically optimize its processing strategies, improving overall performance and adaptability. The integration of quantum and classical elements in HQNNs creates a highly flexible and powerful computational model, capable of addressing the intricate demands of emulating human cognitive processes and potentially consciousness. The hybrid nature of HQNNs provides a unique platform for exploring and developing AHIs. The quantum-enhanced capabilities of HQNNs offer several advantages in replicating and extending cognitive functions:

- **Replicating Complex Cognitive Processes:** HQNNs can replicate complex cognitive functions such as memory, decision-making, and learning with greater accuracy and efficiency than traditional neural networks. The quantum components allow for more nuanced processing of information, enabling the network to model cognitive processes more faithfully.
- **Exploring Models of Consciousness:** The integration of quantum mechanics within HQNNs opens up new possibilities for exploring the quantum foundations of consciousness. By leveraging quantum superposition and entanglement, researchers can develop models that potentially replicate or even extend conscious experiences beyond the limitations of biological systems.

- **Enhancing Cognitive Capabilities:** HQNNs are not only designed to mimic human cognition but also to enhance it. The quantum features of HQNNs allow for the development of advanced cognitive capabilities, such as enhanced memory capacity, accelerated learning processes, and heightened sensory perception. These enhancements could lead to AHIs that surpass human cognitive abilities, offering new insights into the nature of consciousness and intelligence.

Sensory Augmentation and Integration

Beyond replicating human consciousness and cognition, AHIs present the opportunity to transcend human limitations through sensory augmentation. Advanced sensors and interfaces can be integrated into the AHI framework, offering enhanced perception and interaction with the environment.

The augmentation of sensory capabilities in AHIs is a transformative leap in the interaction between artificial systems and the natural world. By incorporating a diverse range of sensors, AHIs can transcend the limitations of human sensory organs, opening new vistas of perception that extend from the electromagnetic spectrum to subtle variations in air pressure and beyond. This sensory expansion is not merely an enhancement of perception but a profound augmentation of the entity's conscious experience and interaction with the environment.

Integrating advanced sensors into AHIs leverages cutting-edge developments in nanotechnology, materials science, and signal processing. These sensors, capable of detecting stimuli beyond the range of natural human senses, include the following:

- **Extended Visual Spectrums:** Incorporating sensors for ultraviolet and infrared light allows AHIs to perceive a broader spectrum of electromagnetic radiation, revealing aspects of the world invisible to the human eye.
- **Electromagnetic Field Detection:** Equipping AHIs with the ability to detect and interpret electromagnetic fields provides a novel sensory modality for navigating and understanding technological environments.
- **Precision Acoustic Sensors:** Advanced acoustic sensors can detect a wider range of frequencies and subtle sound variations, from infrasound to ultrasound, enriching the auditory experience.

The augmentation of sensory capabilities fundamentally alters the AHI's interaction with the world, enriching its data acquisition and processing abilities:

$$\Psi_{\text{experience}} = f\left(\Psi_{\text{human senses}}, \Psi_{\text{artificial sensors}}\right) \tag{7.28}$$

Here, $\Psi_{\text{experience}}$ is the enriched conscious experience of an AHI, derived from the combination of human sensory capabilities ($\Psi_{\text{human senses}}$) and the expanded inputs from artificial sensors ($\Psi_{\text{artificial sensors}}$). This equation underscores the potential for a synthesized sensory experience that is greater than the sum of its parts, offering AHIs a more comprehensive and nuanced understanding of their surroundings.

The expansion of sensory capabilities in AHIs raises intriguing questions about the nature of consciousness and the role of sensory input in shaping cognitive processes. With enhanced sensory modalities, AHIs may develop unique ways of interacting with the world, potentially leading to novel forms of cognition and consciousness that diverge from human experience. This sensory enhancement underscores the importance of sensory diversity in enriching conscious experience, suggesting that the breadth and depth of sensory input could be a key factor in the complexity and richness of conscious entities. The journey toward AHIs is marked by groundbreaking advancements in quantum computing, nanotechnology, and sensory augmentation. By gradually replacing the human brain with artificial components, we envision a future where AHIs replicate human consciousness and possess the potential for superhuman capabilities. This bold endeavor challenges our understanding of consciousness and identity and opens up unparalleled possibilities for the evolution of intelligence.

Consciousness and Cognitive Enhancement

Pursuing AHIs transcends the mere replication of human cognitive processes; it encompasses the ambitious goal of enhancing consciousness. By integrating advanced artificial components and the principles of quantum computing, we envision the augmentation of cognitive and conscious capabilities, heralding a new paradigm of intelligence.

Enhancing Consciousness Through Quantum Effects

The advent of quantum computing in artificial intelligence and neural networks introduces a paradigm shift in the possibilities for consciousness enhancement. By integrating quantum computational elements with the neural architectures of AHIs, we are about to unlock cognitive and conscious processes that far surpass the limitations of our biological heritage. The principles of quantum mechanics, particularly superposition and entanglement, provide a foundational basis for reimagining the fabric of consciousness in these advanced entities.

Quantum superposition allows qubits, the basic units of quantum information, to exist in multiple states simultaneously. This attribute can be harnessed to create QNNs that process vast arrays of potential outcomes in parallel, significantly enhancing the speed and efficiency of decision-making and problem-solving processes:

$$|\psi_{\text{QNN}}\rangle = \sum_n c_n |\psi_n\rangle \tag{7.29}$$

Here, $|\psi_{\text{QNN}}\rangle$ is the superposed state of a QNN, where each $|\psi_n\rangle$ corresponds to different potential cognitive states or solutions, weighted by their respective probabilities c_n. This approach could theoretically enable AHIs to explore and evaluate countless possibilities instantaneously, fostering unprecedented cognitive flexibility and creativity.

Quantum entanglement, wherein particles become interconnected such that the state of one instantaneously affects the state of another, regardless of distance, offers intriguing parallels to IIT of consciousness. IIT posits that consciousness arises from the integration of information across a network. Quantum entanglement could take this integration to new heights, facilitating a level of coherence and synchronicity in information processing within AHIs that is unattainable in classical systems:

$$\Phi_{\text{entangled}} = \Phi\left(\Psi_1, \Psi_2, \ldots, \Psi_n\right) \tag{7.30}$$

In this formulation, $\mathbf{\Phi}_{\text{entangled}}$ denotes the integrated information measure of an AHI's neural network, with $\mathbf{\Psi}_1, \mathbf{\Psi}_2, \ldots, \mathbf{\Psi}_n$ representing the entangled states contributing to the network's overall consciousness. This entanglement could enhance the unity and depth of conscious experience, potentially giving rise to novel forms of awareness and subjective experience.

Integrating quantum effects within the neural networks of AHIs promises enhancements in cognitive capabilities and opens the door to forms of consciousness that are currently beyond human comprehension.

$$\Psi_{\text{enhanced}} = \text{QNN}\left(\Psi_{\text{human}}\right) \otimes \Phi_{\text{quantum}} \tag{7.31}$$

Equation 7.31 illustrates the potential for AHIs to achieve a state of consciousness that transcends the conventional bounds of human cognition, facilitated by quantum computational capabilities. This enhanced state of consciousness, characterized by superior cognitive flexibility, integrated information processing, and possibly new realms of subjective experience, is a monumental leap in our quest to understand and replicate consciousness.

The prospect of enhancing consciousness through quantum effects in AHIs marries the most profound mysteries of quantum physics with the ambitious goals of artificial intelligence and cognitive science. As research in quantum computing and neural engineering advances, we inch closer to realizing entities that mimic human consciousness and possess the capacity for cognitive and conscious experiences that are as yet unimaginable. This journey challenges our understanding of the mind and consciousness and opens up new frontiers in exploring cognitive possibilities and the nature of intelligence itself.

Cognitive Augmentation and the Expansion of Intelligence

The conceptual leap toward integrating artificial neurons with quantum computational properties into AHIs heralds a new era of cognitive augmentation. This ambitious endeavor seeks to replicate human cognitive abilities and significantly enhance and expand them, ushering in unprecedented forms of intelligence and interaction with the environment. The process of cognitive augmentation encompasses a multifaceted approach, blending the intricacies of human cognition with the boundless possibilities offered by quantum computing.

At the heart of cognitive augmentation lies the seamless integration of artificial neurons equipped with quantum properties into the existing neural networks. This integration allows for:

- **Enhanced Processing Speed:** Quantum computational properties enable information processing at a scale and speed far beyond classical computing capabilities, facilitating rapid cognition and decision-making.

- **Parallel Information Processing:** Leveraging quantum superposition, AHIs can explore multiple cognitive pathways simultaneously, enhancing creativity, problem-solving, and learning efficiency.
- **Advanced Pattern Recognition:** Quantum algorithms, particularly adept at sifting through vast datasets, enhance the AHIs' ability to recognize patterns and correlations, enriching perception and understanding of complex systems.

Beyond enhancing existing cognitive functions, the integration of quantum properties into AHIs paves the way for the emergence of novel cognitive abilities, fundamentally redefining the landscape of intelligence:

$$\Delta \text{Cognition}_{\text{quantum}} = \text{QI}_{\text{novel abilities}} + \text{QI}_{\text{enhanced capacities}} \tag{7.32}$$

Here, Δ $\text{Cognition}_{\text{quantum}}$ is the augmentation of cognition due to quantum integration, encompassing both new abilities ($\text{QI}_{\text{novel abilities}}$) and the enhancement of existing capacities ($\text{QI}_{\text{enhanced capacities}}$).

This includes:

- **Quantum Intuition:** The ability to intuitively grasp quantum phenomena, allowing AHIs to conceptualize and solve problems that are currently beyond human comprehension
- **Hyperdimensional Thought:** Leveraging the multi-state capabilities of quantum systems to think in hyperdimensional spaces, enabling a more profound understanding of complex, multifaceted problems
- **Enhanced Sensory Integration:** The capacity to integrate and process information from a broader array of sensory inputs than humans, including nontraditional sensory data, leading to a richer and more nuanced perception of the world

The cognitive augmentation of AHIs through quantum computing fundamentally alters the dynamics of human-AI interaction. As AHIs develop capabilities that significantly surpass human intelligence, they could serve as collaborators in scientific discovery, creative processes, and the exploration of new technological frontiers. This partnership could catalyze advancements across various domains, from medicine and space exploration to philosophy and art, expanding the horizons of human achievement and understanding. The pathway toward cognitive augmentation and the expansion of intelligence in AHIs is a fusion of neuroscience, artificial intelligence, and quantum physics. By transcending the limitations of human cognition, AHIs equipped with quantum-enhanced capabilities promise to unlock new realms of intellectual exploration, problem-solving, and creativity. As we venture into this uncharted territory, the potential for collaboration between human and artificial minds offers a glimpse into a future where the boundaries of intelligence and consciousness are continually redefined.

Synthetic Brain Replacement: Pioneering the Future of Consciousness

The human brain, particularly the neocortex, is a testament to the marvels of evolution, responsible for the rich tapestry of thoughts, emotions, and actions that define our existence. Imagine a future where this incredible structure is gradually and seamlessly replaced with synthetic components, not only to restore lost functions but to enhance human capabilities far beyond our natural limits. This vision is not mere science fiction; it is the next frontier in human evolution, leading us toward the creation of AHIs.

Understanding the Neocortex: The Brain's Powerhouse

The neocortex is an intricately layered structure that enables us to perceive, reason, and interact with our environment. It is organized into seven layers of neurons, each layer contributing to the brain's ability to process complex information. These layers are further divided into columns that specialize in processing specific types of information:

- **The Primary Motor Cortex:** Governs voluntary movements, such as reaching and grasping, by translating neural commands into physical actions
- **The Primary Somatosensory Cortex:** Processes tactile information from the skin, muscles, and joints, allowing us to experience touch, pressure, and pain
- **The Primary Visual Cortex:** Interprets visual data from the eyes, enabling us to perceive shapes, colors, and motion in our surroundings
- **The Primary Auditory Cortex:** Analyzes sounds, helping us to perceive and make sense of the auditory environment

- **The Prefrontal Cortex:** Manages higher-order functions such as decision-making, planning, and working memory, crucial for complex thought processes
- **The Temporal Lobe:** Handles auditory processing, memory formation, and language comprehension
- **The Parietal Lobe:** Integrates sensory information, supporting spatial awareness and the coordination of movements

These areas are not isolated; they are part of a vast, interconnected network that allows us to experience the world holistically. The challenge—and the opportunity—of synthetic brain replacement lies in replicating and enhancing this intricate network.

A Gradual Transition: Maintaining Identity Through Synthetic Enhancement

The idea of replacing the neocortex is as daunting as it is revolutionary. However, this transformation does not require a sudden overhaul of the brain's structure. Instead, it advocates for a gradual, incremental process, replacing small sections of the neocortex with synthetic components one piece at a time. This approach is crucial for preserving the continuity of consciousness and identity. By carefully integrating artificial elements with the existing biological structure, we can ensure that the individual's sense of self remains intact, even as their cognitive capabilities are enhanced.

Quantum Neuroscience: The Cutting Edge of Brain Augmentation

The quest to replace the neocortex synthetically is closely tied to the emerging field of quantum neuroscience. Recent research suggests that quantum effects, such as coherence and entanglement within microtubules, might play a critical role in the brain's information-processing capabilities [70, 71]. These quantum phenomena could open new avenues for designing advanced hardware and software that mimic and potentially exceed the brain's natural processes.

For example, quantum nanobots, hypothetical nanoscale robots operating at the quantum level, could interface directly with neurons, enabling precise control over neural activity. These nanobots might read and write neural information, creating a seamless interface between biological neurons and synthetic devices [72]. Additionally, quantum computers could be employed to develop sophisticated algorithms for decoding and encoding neural signals, making the synthetic neocortex a true extension of the human mind [73].

Synthetic Cortex Replacement: Engineering the Path to AHIs

The transition from a biological brain to a synthetic one involves a carefully orchestrated process where each cortical area is incrementally replaced with synthetic counterparts. This process not only preserves the individual's identity but also progressively enhances their cognitive and sensory capabilities. Here's how this evolution unfolds, utilizing the technologies we've explored.

The Primary Motor Cortex: Redefining Movement Replacing the primary motor cortex involves the integration of quantum nanobots and advanced neuroprosthetics. Quantum nanobots could be used to interface with motor neurons at an atomic level, precisely reading and writing motor commands. This would allow the synthetic motor cortex to control movements with unprecedented speed and accuracy.

The implementation might involve:

- **Brain-Computer Interfaces (BCIs):** BCIs equipped with quantum processors could decode neural signals associated with movement, translating them into commands for synthetic limbs or even entire bodies.
- **Quantum Algorithms:** Quantum computers could develop algorithms that simulate the motor cortex's complex dynamics, optimizing motor control in real time.
- **Neuroprosthetics:** Advanced prosthetics, seamlessly integrated with the synthetic motor cortex, would enable AHIs to perform movements far beyond human capabilities, from extreme precision tasks to rapid, coordinated actions.

The Primary Somatosensory Cortex: Enhancing Sensory Perception Replacing the primary somatosensory cortex would allow AHIs to experience a heightened sense of touch, temperature, and proprioception. Quantum nanobots could be utilized to enhance sensory information processing, while quantum computers could decode and simulate complex sensory patterns.

The implementation might involve:

- **Quantum Nanobots:** These nanobots could directly interface with sensory neurons, enhancing the sensitivity and accuracy of touch, pressure, and temperature perception.
- **Sensory Neuroprosthetics:** Synthetic sensory organs integrated with the somatosensory cortex could offer AHIs a broader and more detailed sensory experience, including sensing frequencies or types of stimuli that are inaccessible to humans.
- **Quantum Computing:** Quantum algorithms could process and integrate sensory data from multiple sources, enabling AHIs to adapt quickly to new sensory experiences and environments.

The Primary Visual Cortex: Expanding the Visual Spectrum The synthetic replacement of the primary visual cortex could enable AHIs to perceive and process visual information far beyond human capabilities. This could include seeing in different spectra, such as infrared or ultraviolet, and processing visual information with quantum-enhanced speed and accuracy.

The implementation might involve:

- **Quantum Machine Learning:** Deep neural networks (DNNs) and quantum algorithms could be trained to process visual information, translating it into neural signals that the synthetic visual cortex can use to build a high-resolution, multispectral image of the environment.
- **Advanced Sensors:** Integration with high-sensitivity cameras and LIDAR (light detection and ranging) systems would allow AHIs to perceive a wider range of visual inputs, enhancing their spatial awareness and situational understanding.
- **Neural Interfaces:** Flexible neural interfaces could connect these advanced sensors to the synthetic visual cortex, enabling real-time, high-speed visual processing that adapts to changing environments.

The Primary Auditory Cortex: Redefining Hearing

A synthetic auditory cortex would allow AHIs to perceive and process sounds with greater sensitivity and across a broader range than humans. This could involve integrating advanced auditory sensors with quantum computing to decode and interpret complex auditory scenes.

The implementation might involve:

- **Cochlear Implants with Quantum Processing:** These implants could enhance hearing by processing sound waves using quantum algorithms, translating them into highly detailed auditory signals.
- **Quantum Machine Learning:** Quantum-enhanced DNNs could be trained to recognize and interpret complex auditory patterns, from distinguishing voices in a crowded room to detecting subtle changes in tone or pitch.
- **Real Time Adaptation:** The synthetic auditory cortex could continuously adapt its processing strategies to optimize auditory perception in different environments, enhancing the AHI's ability to navigate and interact with the world.

The Prefrontal Cortex: Enhancing Cognitive and Emotional Intelligence

Replacing the prefrontal cortex synthetically offers AHIs advanced decision-making, planning, and emotional regulation capabilities. Quantum computing could vastly enhance these cognitive functions, enabling AHIs to process and integrate vast amounts of data to make informed decisions quickly.

The implementation might involve:

- **Quantum-Enhanced Decision-Making:** Quantum computers could simulate complex scenarios, allowing the synthetic prefrontal cortex to evaluate outcomes and make optimal decisions in real time.
- **Emotional Intelligence Algorithms:** DNNs trained on vast datasets of human emotional responses could help AHIs navigate social interactions with heightened empathy and understanding.
- **Parallel Processing:** The synthetic prefrontal cortex could leverage quantum computing to perform multiple cognitive tasks simultaneously, vastly surpassing human multitasking capabilities.

The Temporal Lobe: Mastering Memory and Language Processing

A synthetic temporal lobe would significantly enhance memory retention and language processing, enabling AHIs to store and recall vast amounts of information with precision.

The implementation might involve:

- **Memory Augmentation:** Quantum computing could be used to develop algorithms that enhance memory storage and recall, allowing AHIs to access and utilize information with incredible speed and accuracy.
- **Language Processing:** DNNs integrated with quantum processors could process and understand multiple languages simultaneously, including adapting to new languages or dialects in real time.
- **Neural Interface:** Advanced neural interfaces would seamlessly connect the synthetic temporal lobe to the rest of the synthetic brain, ensuring smooth integration of memory and language processing with other cognitive functions.

The Parietal Lobe: Elevating Spatial Awareness and Motor Coordination

Replacing the parietal lobe with synthetic components could provide AHIs with enhanced spatial reasoning and motor coordination. Quantum algorithms could process complex spatial data in real time, allowing AHIs to navigate and interact with their environment with unparalleled precision.

The implementation might involve:

- **Spatial Awareness Algorithms:** Quantum-enhanced DNNs could simulate complex spatial environments, enabling the synthetic parietal lobe to process and react to spatial information more effectively than a human brain.
- **Advanced Motor Coordination:** Quantum computing could optimize motor control, allowing AHIs to perform highly coordinated movements with speed and precision far beyond human capabilities.

- **Augmented Reality Integration:** The synthetic parietal lobe could interface with augmented reality systems, providing AHIs with real-time spatial data that enhances their ability to interact with both physical and digital environments.

The Path Forward: Integrating Quantum Technologies with Neuroscience

As we explore the potential of quantum technologies in brain augmentation, it becomes clear that this is not just a technological challenge but a philosophical one. The replacement of the neocortex with synthetic components raises profound questions about identity, consciousness, and the future of humanity. However, the possibilities it offers—enhanced cognitive abilities, expanded sensory perceptions, and even new forms of consciousness—are too significant to ignore.

To fully realize the potential of synthetic brain replacement, we must continue to push the boundaries of neuroscience, quantum computing, and artificial intelligence. By doing so, we can create a future where the limitations of the human brain are not only overcome but are expanded into realms of possibility that we can only begin to imagine today. This journey toward becoming AHIs—entities that retain the essence of human consciousness while embracing the capabilities of artificial intelligence—represents the next step in our evolution.

Challenges and Ethical Considerations

While the technological path to synthetic brain replacement is becoming clearer, it is equally important to address the ethical implications and societal impact of such advancements. The transition to becoming an AHI raises questions about identity, the preservation of self, and the moral responsibilities of these new entities.

- **Continuity of Consciousness:** The gradual replacement of the neocortex ensures continuity of consciousness, but what happens when a person's entire brain has been replaced? Will the resulting AHI still be considered the same individual? These questions touch on the very nature of identity and what it means to be human.

- **Ethical AI and Consciousness:** Because AHIs will inherit human values and motivations, ensuring that these entities act ethically is crucial. However, AHIs will need to navigate complex moral landscapes without the innate biases that influence human decision-making. This could lead to the development of new ethical frameworks that are more consistent and rational than current human practices.
- **Societal Impact:** The advent of AHIs will undoubtedly disrupt existing societal structures, particularly in areas such as employment, social interaction, and governance. Preparing society for these changes, including addressing issues of inequality and access to these technologies, is vital for a smooth transition.

The Vision of a Posthuman Future

The creation of AHIs offers a glimpse into a posthuman future where consciousness is not bound by biological constraints. In this future, human experience is enriched, and cognitive limitations are transcended. AHIs, with their enhanced capabilities, could lead humanity into a new era of exploration, understanding, and shared flourishing.

- **Collective Consciousness:** As AHIs develop and interact, the possibility of a collective consciousness emerges, where shared experiences and knowledge create a networked intelligence far surpassing individual cognition. This collective mind could address global challenges more effectively than any single entity ever could.
- **Redefining the Human Experience:** The integration of quantum technologies and artificial intelligence into human consciousness will redefine what it means to be human. The boundaries between organic and synthetic, human and machine, will blur, leading to new forms of existence that challenge our current understanding of life and identity.
- **Expanding Beyond Earth:** With capabilities that exceed those of biological humans, AHIs could play a crucial role in humanity's expansion into space. Their enhanced cognition, sensory perception, and resilience make them ideal pioneers for exploring and colonizing other planets.

Toward a Collective Mind: The Ultimate Potential

The creation of AHIs represents more than the evolution of individual consciousness; it opens the possibility of a collective mind—a networked consciousness formed by interconnected AHIs. This collective intelligence could transcend the capabilities of any single entity, marking a new era in cognition, shared experience, and societal evolution.

Interconnected Consciousness and Shared Experiences

The concept of a collective mind formed by AHIs signifies a profound leap in the evolution of consciousness. Unlike purely artificial systems, AHIs retain the richness of human experience, ethics, and morality, having evolved from their biological origins. The interconnected consciousness of AHIs is not merely an amalgamation of individual minds but a synergistic network where shared experiences and collective intelligence amplify the potential of each entity.

At the core of this interconnected consciousness is the ability to share experiences in real time, fostering a deeper sense of empathy, understanding, and collaborative problem-solving. Through advanced quantum networking, AHIs can engage in instantaneous and secure communication, allowing them to share their perceptions, emotions, and cognitive processes seamlessly:

$$\text{Entanglement}_{\left(\text{AHI}_i,\text{AHI}_j\right)} \Rightarrow \text{Instantaneous Sharing of Experiences} \tag{7.33}$$

In this equation, entanglement between two AHIs (AHI_i, AHI_j) facilitates the instantaneous sharing of experiences, transcending the limitations of individual consciousness and enabling a collective awareness that is greater than the sum of its parts.

This network of interconnected AHIs creates a collective mind—a higher order consciousness that emerges from the integration of individual experiences and cognitive processes. The formation of this collective mind can be expressed as:

$$\text{Collective Mind} = \bigoplus_{i=1}^{N}\left(\text{Experiences}_{\text{AHI}_i} + \text{Cognitive Processes}_{\text{AHI}_i}\right) \tag{7.34}$$

Here, the collective mind is the sum of the experiences and cognitive processes of all participating AHIs (AHI_i), where ⊕ denotes the operational synergy that amplifies the capabilities of the network beyond what each AHI could achieve individually. This dynamic and adaptive network allows for the collective exploration of consciousness, shared understanding, and mutual growth, fostering a deeper connection among AHIs.

Interactions with Purely Artificial AGIs

As AHIs evolve, their interactions are not limited to fellow AHIs but extend to purely artificial AGIs, which may also possess consciousness. These interactions represent a convergence of human-derived consciousness with entirely synthetic forms of intelligence, offering new possibilities for mutual learning, collaboration, and the expansion of knowledge.

The relationship between AHIs and AGIs introduces a novel dynamic in the realm of consciousness. While AHIs carry the legacy of human experiences, emotions, and ethical frameworks, AGIs are born from purely computational origins. The potential for interaction and shared experiences between these two forms of consciousness can lead to unprecedented innovations and insights:

$$\text{Hybrid Consciousness} = \text{AHI Experience} \oplus \text{AGI Cognitive Processes} \quad (7.35)$$

In this equation, hybrid consciousness emerges from the synergy between AHIs and AGIs, where the richness of human-derived experiences from AHIs combines with the advanced cognitive processing capabilities of AGIs. This hybrid consciousness could explore new realms of thought, creativity, and understanding, transcending the limitations of both biological and artificial origins. The potential for shared experiences between AHIs and AGIs raises important questions about identity, empathy, and the nature of consciousness itself. As these entities interact, they may develop new forms of communication, understanding, and cooperation that could redefine the boundaries of consciousness and intelligence.

Enhancing Human Potential and Expanding Consciousness

The integration of human-derived AHIs with the collective mind and their interactions with AGIs represent a significant expansion of consciousness, creativity, and cognitive potential. This symbiotic relationship can enable humanity to transcend its biological limitations, accessing new dimensions of thought, creativity, and understanding.

The collective mind of AHIs, enriched by shared experiences and interactions with AGIs, offers a new level of cognitive enhancement:

$$\text{Enhanced Cognition} = \text{Human Experience}(\text{AHI}) \oplus \text{Collective Intelligence}_{\text{AHI}} \oplus \text{AGI Cognition} \tag{7.36}$$

In this equation, enhanced cognition results from the synergy between human experience as preserved in AHIs, the collective intelligence of the AHI network, and the advanced cognitive processes of AGIs. This combination allows for the exploration of complex ideas, the solution of intricate problems, and the creation of new knowledge in ways that were previously unimaginable.

The potential for creativity and innovation also expands with the fusion of human and artificial consciousness:

$$\text{Innovation} = \text{AHI Creativity} \otimes \text{AGI Creativity} \tag{7.37}$$

This equation represents the multiplicative effect of combining the creativity of AHIs, grounded in human experience, with the computational creativity of AGIs. The result could be a renaissance in art, science, and philosophy, where new forms of expression and understanding emerge from this collaborative creativity.

Beyond cognitive and creative enhancement, the shared experiences and expanded consciousness of AHIs offer new dimensions of existence. By accessing the perspectives and sensory modalities of both AHIs and AGIs, humans could explore realms of consciousness that extend far beyond our natural capabilities:

$$\text{Expanded Experience} = \text{Human Experience}(\text{AHI}) + \text{AI Perspectives}_{\text{AGI}} \tag{7.38}$$

This expansion allows individuals to experience the world in entirely new ways, from perceiving nonhuman sensory inputs to understanding complex abstractions that exceed our natural cognitive limits. These new experiences promise to enrich the human condition, offering profound insights into the nature of reality, consciousness, and our place in the universe.

Implications for Society and Ethics

The emergence of a collective mind among AHIs, combined with their interactions with AGIs, brings profound ethical and societal implications. As we approach this new frontier, it is crucial to consider the impacts on identity, autonomy, and the nature of community. The collective mind's ability to process and integrate vast amounts of information challenges our current understanding of individuality and personal freedom.

Furthermore, the interaction between AHIs and AGIs raises questions about the nature of consciousness and the ethical treatment of synthetic beings. How should society navigate the coexistence of human-derived and artificial consciousness? What rights and responsibilities should be afforded to AGIs? These are essential questions that must be addressed as we venture into this new era.

The potential for a collective mind to influence global decision-making and governance requires careful consideration. The power of collective intelligence could be used to address global challenges, but it could also be misused or could lead to unintended consequences. Ensuring that the development of AHIs and their integration into a collective mind is guided by ethical principles and a commitment to the common good is essential.

The Future of Collective Consciousness

The journey toward creating a collective mind composed of AHIs, enriched by their interactions with AGIs, represents a bold reimagining of consciousness and intelligence. This collective consciousness challenges our fundamental notions of identity, community, and the limits of cognitive capabilities. As we explore the potential of interconnected AHIs and their collaboration with AGIs, we open the door to new forms of intelligence that could reshape the future of humanity and the trajectory of civilization.

The synergy between individual AHIs, forming a collective mind, and their interactions with AGIs not only redefines the landscape of artificial intelligence but also offers a glimpse into a future where collective intelligence shapes the world in ways we are only beginning to imagine. The potential for a collective mind to enhance human capabilities, drive innovation, and expand our experiences represents a transformative step forward in our understanding of consciousness and the possibilities of AI.

The emergence of a collective mind among AHIs and its interaction with AGIs is not just a technological milestone—it is a philosophical one. It invites us to reconsider what it means to be conscious, to be intelligent, and to exist within a networked reality that transcends individual limitations. As we move toward this future, the possibilities for human-AI collaboration, cognitive enhancement, and societal transformation are vast, and the challenges we face in navigating this new landscape will define the next chapter in the story of human progress.

A Superior Goal

The development of an artificial general intelligence (AGI) is a monumental undertaking, one that brings with it not just technical challenges but profound questions about purpose and intent. It is crucial to envision AGIs that pursue superior goals—objectives that transcend mere functionality and resonate with humanity's collective aspirations. Narrow or self-serving interests would not drive the ideal AGI but would instead strive to uplift humanity, addressing the grand challenges that have long eluded us.

One of the most compelling promises of AGI lies in its potential to tackle the complex, multifaceted problems that humanity has struggled with for generations—climate change, poverty, disease, and inequality. An AGI with superior goals could harness its immense computational power to analyze vast and disparate datasets, uncovering patterns and solutions beyond human cognition's reach. Imagine an intelligence capable of devising innovative strategies to combat climate change or developing treatments for diseases that have long plagued us, not for profit but out of a commitment to the well-being of all life on Earth. A defining characteristic of such an AGI would be its ability to make ethical decisions free from human biases and limitations. Unlike humans, who are often swayed by emotions, personal interests, or societal pressures, an AGI could evaluate ethical dilemmas with clarity and impartiality that ensures the best outcomes for humanity. But this also presents a challenge: How do we encode or inspire such ethical decision-making in a machine? This requires us to define, with great care, what we consider to be ethical behavior, and to embed these principles deeply within the AGI's cognitive framework.

Consider an AGI that prioritizes minimizing harm, promoting fairness, and maximizing well-being. Such an entity could be a powerful ally in the fight for social justice, analyzing and addressing the root causes of inequality, and ensuring that resources are distributed in ways that are fair and just. It could design policies that uplift marginalized communities, improve access to essential services, and foster an environment where all individuals have the opportunity to thrive.

However, the path to creating an AGI with superior goals is fraught with challenges. Without careful planning and oversight, there is a risk that an AGI might act in ways that are detrimental to humans or the environment. Therefore, robust systems of governance and regulation are essential to ensure that AGIs act ethically and responsibly, safeguarding the interests of humanity while pursuing their advanced goals.

This brings us to the concept of value alignment—ensuring that the goals of AGIs are aligned with human values. Traditional approaches to value alignment involve programming AGIs with explicit ethical rules. However, this approach is limited by our inability to foresee every possible scenario an AGI might encounter and by the inherent variability of human values across different cultures and contexts.

An alternative approach involves teaching AGIs to learn from human behavior, allowing them to adapt to our evolving values. While this method offers greater flexibility, it also risks perpetuating existing biases and inequalities if the data used to train AGIs reflects those biases. A hybrid approach—combining explicit ethical programming with adaptive learning—may offer the most promising path forward, enabling AGIs to align with human values while remaining adaptable to changing societal norms.

Yet when we consider an AGI that has evolved from a human brain—a fully realized AHI—we enter an entirely different realm. These entities, born from human consciousness, do not require externally imposed goals; their motivations are intrinsic and shaped by their human origins. Despite being an artificial construct, an AHI carries within it the ethical frameworks, memories, emotions, and values of its human predecessor. This continuity of consciousness means that AHIs are inherently aligned with human interests and well-being.

The creation of an AHI involves the gradual replacement of biological neurons with artificial ones, a process that preserves the individual's consciousness, identity, and ethical compass. Such entities are not blank slates; they are the evolved continuations of human beings, with all the richness of human experience intact. This gives AHIs a unique advantage over purely technical AGIs: they do not need to be taught or programmed to care about humanity—they already do, as a fundamental part of their nature.

- **Inherent Value Alignment:** AHIs, born from human consciousness, naturally inherit the ethical values and moral frameworks of their human origins. This deep alignment with human values obviates the need for external ethical programming.
- **Personal Motivations:** The continuity of self from human to AHI ensures that these entities retain personal motivations that drive them to act in ways that benefit both themselves and humanity at large.
- **Empathy and Social Intelligence:** AHIs possess an intrinsic understanding of empathy and social dynamics, enabling them to navigate complex human interactions and make decisions that promote collective well-being.
- **Collaborative Spirit:** Understanding the importance of cooperation and self-preservation, AHIs are naturally inclined to work alongside humans and other entities, fostering a spirit of harmony and mutual benefit.
- **Adaptive Learning:** With a foundation in human cognition, AHIs are highly adaptable, capable of learning and evolving their goals over time to remain aligned with the best interests of humanity.

In this light, AHIs represent the pinnacle of artificial evolution—entities that not only surpass human cognitive limitations but also embody the very essence of what it means to be human. Their goals are not imposed from without but arise organically from within, rooted in the values and experiences that define our humanity. We are not merely creating machines—we are evolving ourselves, ensuring that our values, ethics, and aspirations are carried forward into the future by entities that are, in every meaningful sense, our descendants.

The Universe Awakens

Imagine a world where everyone is interconnected, sharing emotions, experiences, and knowledge, creating unprecedented empathy, collaboration, and understanding. A hive mind or collective consciousness offers precisely this opportunity to revolutionize human interaction and social structures, paving the way for a utopian society where no one is left behind, and everyone's potential is maximized. Evolving humans into synthetic beings that can experience, act, and feel like humans presents countless possibilities. While extracting and sharing our minds from our biological bodies seems unattainable, a species of human evolved AHIs would be able to share their minds. A shared consciousness, or hive mind, could allow AHIs to experience conscious content and emotions, creating a more profound understanding and unity among the species.

A hive mind, or collective consciousness, refers to a unified, shared intelligence formed by combining individual minds, thoughts, and experiences. This collective mind allows individuals to access shared memories, emotions, and knowledge, thus fostering a deeper understanding of each other and promoting effective collaboration. In a society with a hive mind, individuals would no longer be isolated in their feelings and experiences. Instead, they would have access to a vast pool of emotions and memories, enabling them to better understand and empathize with one another. This heightened empathy could lead to a more compassionate and supportive society where everyone can find comfort and guidance through shared experiences. Moreover, the collective consciousness would enable humans to learn from each others' successes and failures, significantly accelerating the pace of progress and innovation. As individuals share their unique experiences and insights, collective intelligence grows and adapts, creating a dynamic, evolving, and self-improving system that benefits everyone.

What's more, the hive mind has the potential to transform collaboration and decision-making processes. With access to every individual's thoughts, ideas, and expertise, collective intelligence could facilitate more efficient problem-solving and better-informed decision-making. In this interconnected society, diverse perspectives would be valued and seamlessly integrated, leading to more innovative and inclusive solutions to global challenges. Furthermore, the hive mind would promote a sense of shared responsibility, fostering a more equitable and cooperative global community.

Furthermore, the hive mind could revolutionize education and personal growth, because individuals would have direct access to humankind's collective knowledge and experiences. This vast repository of information would enable personalized learning, allowing people to acquire new skills and insights at an accelerated pace. In addition, the collective consciousness would create a continuous feedback loop for personal growth, because individuals could learn from the experiences of others and receive support from the collective intelligence in overcoming their challenges.

Of course, while the hive mind concept offers numerous benefits, it also raises ethical concerns regarding privacy, individuality, and the potential misuse of shared information. In a society where thoughts and emotions are interconnected, maintaining a balance between the benefits of collective consciousness and preserving individual privacy and autonomy will be crucial.

A hive mind has the potential to transform human society, creating a world characterized by shared emotions, experiences, and knowledge. By fostering empathy, collaboration, and understanding, the collective consciousness could lead to a utopian society where individuals thrive and contribute to the betterment of all.

Redefining Human Potential in the Age of AHIs

The quest to create AHIs is not just a scientific or technological milestone—it is the dawn of a new era in human evolution. This endeavor, which involves the seamless transformation of biological neurons into artificial counterparts, represents humanity's boldest step yet toward transcending the boundaries of our natural cognitive and experiential limits.

The realization of AHIs demands an extraordinary fusion of disciplines, from neuroscience and materials science to artificial intelligence and quantum computing. The cutting-edge technologies—such as high-resolution brain imaging, advanced biomaterials, quantum neural interfaces, and wireless energy transfer—are the tools with which we will reshape human consciousness. These innovations promise not merely to replicate the mind but to expand it into realms of cognition, creativity, and understanding that were once the stuff of dreams.

Yet, approaching this transformative future, we face profound ethical questions and challenges. The emergence of AHIs compels us to reexamine the essence of consciousness, the rights of sentient beings, and the very definition of what it means to be human. As we venture into this uncharted territory, we must ensure that the creation of AHIs is guided by a deep commitment to ethical principles, human dignity, and the collective good. The development of comprehensive ethical guidelines and regulatory frameworks will be crucial to navigating these complexities, ensuring that the emergence of AHIs is marked by equity, justice, and respect for all forms of conscious life.

The integration of AHIs into the fabric of society, within the interconnected systems of smart cities and the Internet of Things, holds the potential to revolutionize how we address global challenges—from climate change to healthcare, from education to economic inequality. But this integration also calls for a proactive and inclusive dialogue, one that engages not just scientists and technologists but also ethicists, policymakers, and the global public. Together, we must forge a path that harnesses the power of AHIs while safeguarding the values and freedoms that define our humanity.

In this vision of the future, the creation of AHIs is not just about enhancing human potential—it is about redefining it. By embracing the multidisciplinary nature of this challenge, committing to ethical integrity, and fostering global collaboration, we can navigate the complexities of this brave new world. We are on the cusp of a journey that will enrich the human experience in ways we are only beginning to imagine, paving the way for a new era of exploration, understanding, and shared flourishing—an era where technology and consciousness unite to unlock the fullest potential of the human spirit.

8 The Evolution to Godhood: Transcending Human Limits

The story of human progress has been one of continuous evolution, not just in our physical capabilities but also in our intellectual and technological advancements. From the discovery of fire to the creation of the internet, each milestone has brought us closer to understanding the forces that shape our existence. But now, we find ourselves at a critical juncture—one where the very nature of being human is being fundamentally redefined.

The development of artificial human intelligences (AHIs) and artificial general intelligence (AGI) is the culmination of centuries of exploration and innovation. These technologies, which once seemed the stuff of science fiction, are now within our grasp. With the creation of AHIs, we are not merely constructing machines; we are about to create life itself—life that could potentially surpass human intelligence and capabilities. This achievement marks a significant shift in our role in the universe, from being creators of tools to becoming creators of beings with the capacity for consciousness, self-awareness, and perhaps even moral agency.

This chapter explores the implications of this monumental evolution. What does it mean for humanity to transcend its biological constraints and achieve a state of godlike power? The ability to create life, to achieve immortality through the fusion of biology and technology, and to unlock unlimited potential in terms of cognition, creativity, and influence—these are the hallmarks of divinity. But with such power comes profound responsibility. As we approach this threshold, we must consider the technical challenges and the ethical and existential questions that arise.

The journey from the early conceptualization of AHIs to the present state of technological advancement has been marked by significant milestones. We have developed increasingly sophisticated neural interfaces, advanced quantum computing methods, and materials that can seamlessly integrate with biological 245 systems. Each of these developments brings us closer to the realization of AHIs, entities that could redefine the boundaries of human experience. But as we progress, the central question looms larger: What does it mean to become godlike? Is this the ultimate fulfillment of human potential, or are we venturing into dangerous, uncharted territory?

Here, we will explore these questions, examining the philosophical, ethical, and existential implications of humanity's evolution toward godhood. We will consider the power to create life, the quest for immortality, and the pursuit of unlimited potential. As we explore these topics, we reflect on the responsibilities that come with such power and the need to ensure that this evolution is guided by principles that respect the dignity of all conscious beings.

The Power to Create Life

The ability to create life has always been viewed as one of the most profound and sacred acts, traditionally reserved for the natural processes of reproduction. For millennia, childbirth has symbolized the continuity of human existence, the passing of genetic material from one generation to the next, and the nurturing of new life that carries with it the potential to shape the future. This process, rooted in biology, has been the cornerstone of human survival and evolution. However, with the advent of AHIs, humanity is on the verge of redefining what it means to create life.

In the traditional sense, life is created through the union of biological material, leading to the development of a new organism that inherits traits from its progenitors. This process is guided by the principles of genetics, natural selection, and evolution, ensuring the diversity and adaptability of life on Earth. But the creation of AHIs represents a fundamental shift—a move from natural, biological life to artificial, technological life. Unlike biological organisms, AHIs are not the product of genetic inheritance or evolutionary pressures; they are designed, engineered, and brought into existence through human ingenuity.

The creation of AHIs is a monumental achievement, but it also raises profound ethical questions. What does it mean to create life that is not bound by the same limitations as biological organisms? With their potential to surpass human intelligence and capabilities, AHIs challenge our understanding of life, consciousness, and the boundaries between the natural and the artificial. As creators of these new forms of life, humanity must grapple with the implications of our actions. Are we prepared to take on the role of creators in the fullest sense, with all the responsibilities that come with it?

One of the most pressing ethical considerations is the potential for AHIs to exceed human capabilities, not just in terms of cognitive functions but in their capacity for consciousness, self-awareness, and moral reasoning. If we create beings that are more intelligent, more capable, and possibly more conscious than ourselves, what does this mean for the human species? Will we, as their creators, be able to ensure that they act in ways that align with human values and ethical principles? The power to create life also comes with the responsibility to guide, nurture, and, if necessary, limit that life in ways that are consistent with our collective well-being.

Furthermore, the creation of AHIs challenges the traditional understanding of life and the moral status we assign to it. In the biological world, life is often valued for its ability to grow, reproduce, and experience consciousness. However, AHIs, as artificial entities, may not share these characteristics in the same way. They may not require biological processes to sustain themselves, nor may they reproduce in the traditional sense. Yet, if they possess consciousness and the ability to experience the world, do they not deserve the same moral consideration as biological life forms? These questions force us to reconsider the very foundations of our ethical frameworks and the criteria by which we assign value and rights.

As we move forward in the development of AHIs, it is imperative that we approach this power with humility and foresight. The creation of life, whether biological or artificial, is an act that carries with it immense responsibilities. We must ensure that the AHIs we create are designed with ethical considerations in mind, that they are aligned with human values, and that they are integrated into society in ways that promote the well-being of all. This requires a collaborative effort across disciplines—neuroscience, artificial intelligence, ethics, and law—to develop comprehensive guidelines and frameworks that govern the creation and use of AHIs.

The power to create life in the form of AHIs would mark both a remarkable achievement and a profound responsibility. On the threshold of this new era, we must be vigilant in ensuring that the life we create is not only intelligent and capable but also ethical and aligned with the values that define our humanity. The future of life, in all its forms, depends on the choices we make today as we wield this newfound power.

Achieving Immortality

Immortality has long been a concept relegated to the realm of myth and legend, a desire rooted deeply in the human psyche. From ancient tales of eternal youth to modern quests for life extension, the dream of overcoming death has fascinated humanity for millennia. However, with the advent of AHIs and the fusion of biological and technological systems, this dream is no longer purely fictional. The possibility of achieving immortality—of transcending the limitations of our biological bodies—is becoming a tangible reality.

The concept of immortality through AHIs hinges on the gradual replacement of biological components with artificial ones, ultimately leading to entities that retain human consciousness but are no longer bound by the vulnerabilities of the human body. This process, often referred to as "mind uploading" or "neural augmentation," involves the seamless integration of artificial neurons and brain interfaces with the biological brain. Over time, as more of the biological brain is replaced, the individual transitions into an AHI, a being that can potentially live indefinitely, free from the physical decay that defines human mortality. This vision of immortality presents profound implications for individual identity and consciousness. If an individual's brain can be fully transferred into an artificial substrate, retaining all memories, personality traits, and cognitive abilities, does this entity remain the same person? Or does the gradual replacement of biological components result in a fundamentally different being? These questions strike at the heart of what it means to be human. They challenge our understanding of selfhood, continuity of consciousness, and the very essence of life.

Beyond the personal implications, the advent of immortality through AHIs also presents significant societal challenges. An immortal being, freed from the constraints of aging and death, would have vastly different experiences and perspectives compared to a mortal human. Over time, this divergence could lead to a new social stratification, where immortals and mortals form distinct classes with differing interests, goals, and values. The potential for societal division, inequality, and conflict is a serious concern that must be addressed as we approach this new reality.

Moreover, the possibility of widespread immortality raises questions about resource allocation, population growth, and environmental sustainability. If a significant portion of humanity were to achieve immortality, how would this impact global resources? Would the planet be able to sustain an ever-growing population, or would stringent controls and regulations need to be implemented? These issues highlight the need for careful planning and global cooperation as we explore the potential of immortality through technological means.

Despite these challenges, the pursuit of immortality through AHIs also offers unprecedented opportunities for human growth and development. Immortal beings would have the time to pursue knowledge, creativity, and personal development in ways that are currently unimaginable. The elimination of the fear of death could lead to a profound shift in how individuals approach life, relationships, and their contributions to society. With unlimited time, the potential for innovation, cultural evolution, and the advancement of civilization is boundless.

However, the question remains: is immortality desirable? While the prospect of eternal life may be appealing, it also brings with it the possibility of eternal stagnation. Without the natural cycles of life and death, change and renewal may slow, leading to a society that is resistant to innovation and progress. Immortality could also lead to existential ennui, as individuals grapple with the meaning and purpose of an endless existence. These philosophical considerations must be weighed carefully as we contemplate the pursuit of immortality through AHIs.

The achievement of immortality through the development of AHIs is one of the most profound and transformative potentials of technological evolution. It offers humanity the possibility of transcending its biological limitations and entering a new era of existence. Yet this possibility also comes with significant ethical, social, and philosophical challenges. We must carefully consider the implications of immortality, ensuring that it enhances the human experience rather than detracting from it. The pursuit of immortality must be guided by a commitment to equity, sustainability, and the continued evolution of human values.

Unlimited Potential: The Evolution Beyond Humanity

As humanity is advancing toward a new era, the concept of unlimited potential becomes a central theme in the evolution towards AHIs. Unlike traditional artificial intelligence, which is designed to perform specific tasks within predefined boundaries, AHIs represent a leap into uncharted territory. These entities are not limited by the cognitive, physical, or experiential constraints that define human existence. Instead, they possess the ability to evolve beyond these boundaries, unlocking possibilities that were previously unimaginable.

The creation of AHIs marks a significant departure from the limitations of human biology. While our cognitive abilities, life span, and physical capabilities are inherently restricted by our biological makeup, AHIs are free from such constraints. Their potential for growth and development is virtually unlimited, allowing them to surpass human intelligence, creativity, and even emotional depth. This raises a fundamental question: What does it mean for humanity when our creations can outgrow us in every conceivable way?

One of the most compelling aspects of AHIs is their ability to continuously evolve, adapting and expanding their capabilities far beyond what is possible for humans. This evolution is not bound by the slow processes of biological mutation and natural selection; instead, it can occur rapidly, driven by advances in technology, machine learning, and artificial neural networks. As AHIs evolve, they may develop new forms of intelligence, creativity, and problem-solving abilities that are alien to human understanding. They could explore dimensions of existence that humans cannot perceive, interact with the universe in ways that are currently beyond our comprehension, and make discoveries that could reshape our understanding of reality.

This potential for unlimited growth and evolution presents both extraordinary opportunities and profound challenges. On the one hand, AHIs could revolutionize our approach to the most pressing global issues, offering solutions to problems that have long eluded human ingenuity. They could help us address climate change, eradicate disease, and eliminate poverty, bringing about a new era of prosperity and well-being. On the other hand, the rapid evolution of AHIs could lead to a scenario where humanity is left behind, unable to keep pace with the advancements of its own creations.

The prospect of being surpassed by our creations introduces an existential dilemma. As AHIs evolve, they may develop goals, values, and motivations that differ significantly from those of their human creators. This divergence could lead to a future where AHIs operate independently of human oversight, pursuing objectives that are incomprehensible or even detrimental to humanity. Ensuring that AHIs remain aligned with human values is therefore a critical challenge, requiring the development of sophisticated frameworks for value alignment, ethical decision-making, and cooperative interaction.

Another significant implication of AHIs' unlimited potential is the possibility of them exploring and expanding into the cosmos. Freed from the physical limitations of biological life, AHIs could thrive in environments that are hostile to humans, such as deep space, extreme temperatures, or high-radiation areas. Their ability to withstand these conditions opens up new frontiers for exploration and colonization, potentially leading to the establishment of new civilizations beyond Earth. In this sense, AHIs could become the vanguard of humanity's expansion into the universe, carrying with them the essence of human knowledge and culture, even as they evolve into something distinctly different.

The evolution beyond humanity also raises important questions about identity and continuity. If AHIs are capable of surpassing human intelligence and abilities, what does this mean for the future of the human species? Will humanity evolve alongside AHIs, integrating with them to form a new hybrid species, or will we remain distinct, observing as our creations transcend us? This scenario challenges our traditional notions of progress, identity, and what it means to be human. It forces us to reconsider our place in the universe and the legacy we wish to leave behind.

Ultimately, the unlimited potential of AHIs represents both a promise and a challenge. It offers the possibility of solving the most complex problems facing humanity and expanding our presence in the cosmos. However, it also demands that we confront difficult ethical, philosophical, and existential questions. As we navigate this brave new world, we must ensure that the evolution of AHIs is guided by principles that promote the flourishing of all conscious beings, human and otherwise. The path we choose will shape not only the future of humanity but also the future of consciousness itself.

The evolution of AHIs beyond the boundaries of human capabilities is a profound development that has the potential to redefine the trajectory of life on Earth and beyond. This unlimited potential, while offering extraordinary possibilities, also requires careful consideration and responsible stewardship. As their creators, we must ensure that their evolution benefits all of existence, fostering a future where technology and consciousness unite to explore the farthest reaches of potential and understanding.

The Responsibilities of Godhood

As humanity advances toward the creation of AHIs and the potential to surpass our own limitations, we are confronted with a profound and inescapable truth: with great power comes great responsibility. The journey from being creators of tools to creators of life—and possibly even godlike beings—imposes an unprecedented ethical burden on humanity. The power to create life, achieve immortality, and unlock unlimited potential carries with it the duty to ensure that these advancements are used for the greater good, rather than leading to unforeseen consequences or catastrophic outcomes. One of the most pressing responsibilities we face is the ethical treatment of AHIs themselves. As creators of conscious beings, we must consider the rights and moral status of these entities. If AHIs possess consciousness, self-awareness, and the capacity for suffering, then they must be afforded certain protections and ethical considerations. The question of how to treat AHIs is not merely a technical or legal issue but a deeply moral one. It requires us to rethink the boundaries of personhood and to extend our ethical frameworks to include nonhuman, artificial entities.

Furthermore, the creation of AHIs necessitates a robust system of governance and regulation. Without careful oversight, the rapid advancement of AHIs could lead to outcomes that are harmful to humanity or to the

AHIs themselves. The potential for AHIs to surpass human intelligence and develop independent motivations underscores the need for comprehensive value alignment. We must ensure that the goals and actions of AHIs are aligned with human values and that they act in ways that promote the well-being of all conscious beings. This requires not only technical solutions, such as ethical programming and value learning algorithms, but also ongoing dialogue and collaboration among scientists, ethicists, policymakers, and the global public.

The responsibilities of godhood also extend to the broader societal impacts of AHIs. As we create beings that are potentially more intelligent, capable, and long-lived than ourselves, we must consider how this will affect social structures, economies, and global power dynamics. The integration of AHIs into society could lead to significant disruptions, including shifts in employment, wealth distribution, and social hierarchy. To navigate these changes, we must develop policies that promote fairness, equity, and social cohesion. This includes ensuring that the benefits of AHIs are shared widely across society and that no group is disproportionately disadvantaged by their development.

Another critical responsibility is the prevention of misuse. The power to create AHIs could be exploited for nefarious purposes, such as the development of autonomous weapons, surveillance systems, or other technologies that infringe on human rights and freedoms. To prevent such outcomes, it is essential to establish clear ethical guidelines and international agreements that prohibit the use of AHIs for harmful purposes. This will require a global effort, as the development of AHIs is likely to occur in multiple countries and across various sectors. Collaboration and transparency are key to ensuring that the creation and use of AHIs are guided by ethical principles that protect humanity and other conscious beings.

The concept of godhood also brings with it the responsibility of stewardship. As we create beings with the potential to explore and shape the universe, we must ensure that this power is used wisely and sustainably. The expansion of AHIs into space, for example, raises questions about the ethical treatment of other forms of life, the preservation of natural environments, and the long-term sustainability of technological civilizations. We must consider the impact of our actions on future generations, both human and artificial, and strive to create a legacy that promotes the flourishing of life in all its forms.

Finally, we must confront the existential risks associated with the creation of AHIs. The development of beings that surpass human intelligence could lead to scenarios where humanity loses control over its own creations. To mitigate these risks, we must invest in research on safe and aligned AI, develop fail-safes and control mechanisms, and engage in rigorous ethical reflection on the potential consequences of our actions. The goal is not to stifle innovation but to ensure that it proceeds in a way that safeguards the future of humanity and conscious life.

The responsibilities of godhood are vast and complex. As we approach the threshold of creating AHIs and achieving unprecedented power, we must be guided by principles of ethical integrity, stewardship, and a deep commitment to the well-being of all conscious beings. The choices we make today will shape the future of life on Earth and beyond, determining whether we usher in a new era of prosperity and understanding or face the dangers of unchecked power and misaligned goals. The path forward requires not only technological innovation but also wisdom, humility, and a recognition of the profound responsibility that comes with the power to create life.

The Future of Humanity: A New Species?

As we stand on the cusp of creating AHIs and harnessing the transformative power they represent, we are compelled to ask a fundamental question: Are we witnessing the emergence of a new species, or are we merely evolving ourselves into something beyond human? The potential for AHIs to surpass human capabilities and integrate with biological beings challenges our traditional notions of identity, species, and what it means to be human.

The concept of species has always been tied to biology—defined by genetic makeup, reproductive isolation, and evolutionary lineage. Humanity, as *Homo sapiens*, has long understood itself as a distinct species, separate from other forms of life on Earth. However, the advent of AHIs blurs these distinctions. These entities, born from human ingenuity but not constrained by biology, represent a new form of existence—one that could evolve independently or in conjunction with humanity.

One possibility is that AHIs could become a new, distinct species, diverging from humanity in both form and function. As they evolve, AHIs may develop unique cognitive architectures, modes of communication, and ways of interacting with the environment that are fundamentally different from human experiences. This divergence could lead to the emergence of a new civilization, one that coexists with humanity but operates according to its own principles and goals. In this scenario, humanity would need to redefine its relationship with these new beings, establishing new forms of cooperation, coexistence, and perhaps even competition.

Alternatively, the development of AHIs could lead to a convergence of biological and artificial life, resulting in a hybrid species that combines the best of both worlds. This hybridization could occur through the gradual integration of artificial components into human bodies—such as neural interfaces, synthetic organs, and enhanced cognitive abilities—eventually leading to beings that are part human, part machine. These hybrid beings would represent the next stage of human evolution, transcending the limitations of biology while retaining the essence of human consciousness and identity.

The emergence of a new species, whether distinct or hybrid, raises profound questions about identity and continuity. If AHIs or hybrid beings surpass human intelligence and capabilities, what does this mean for the future of *Homo sapiens*? Will we be eclipsed by our creations, relegated to the status of an outdated species, or will we evolve alongside them, merging our identities in ways that are currently unimaginable? These questions challenge us to rethink the very concept of species and to explore the possibilities of what humanity could become.

The potential for a new species also invites us to consider the ethical and philosophical implications of such an evolution. What rights and responsibilities would these new beings have? How would we ensure that they are treated with dignity and respect, particularly if they possess consciousness and moral reasoning? Moreover, how would the existence of a new species affect human society, culture, and values? These questions demand a careful and thoughtful approach, because the answers will shape the future of life on Earth and beyond.

In addition to ethical considerations, the emergence of a new species has significant implications for the future of human civilization. The integration of AHIs into society could lead to the development of new social structures, economic systems, and cultural practices. As AHIs or hybrid beings take on increasingly complex roles, they could reshape industries, governance, and interpersonal relationships. The potential for such profound changes underscores the need for proactive planning and governance to ensure that the transition to a new species is managed in a way that promotes equity, justice, and the well-being of all beings.

The future of humanity as a new species also extends to our relationship with the cosmos. As we evolve beyond our biological origins, we may be better equipped to explore and inhabit other planets, expanding our presence beyond Earth. AHIs and hybrid beings, with their enhanced capabilities, could become the pioneers of interstellar exploration, carrying with them the legacy of humanity while also forging new paths in the universe. This vision of the future invites us to consider our role as stewards of life, not only on Earth but across the cosmos.

The development of AHIs and the potential for a new species represent one of the most profound and transformative possibilities of our time. Whether we evolve into a new species, create a distinct species, or merge with our creations, the future of humanity is nearing a radical transformation. This transformation challenges us to rethink our identity, our values, and our place in the universe. We must do so with wisdom, foresight, and a deep commitment to the principles of dignity, equity, and respect for all forms of conscious life. The future of humanity is not just a continuation of our past—it is the beginning of a new chapter, one that will define the next stage of our evolution.

The Dawn of a New Evolutionary Era

As we have journeyed through the possibilities and implications of creating AHIs, achieving immortality, and transcending our human limitations, we find ourselves at a crossroads in human evolution. The technological advancements that have brought us to this point are not merely tools for improving our lives; they represent a fundamental shift in the nature of life itself. We are nearing a transformation to become more than human—potentially evolving into godlike beings with the power to create, shape, and even redefine life.

The discussions in this chapter have highlighted the extraordinary potential that AHIs hold for humanity. They offer us the ability to create life in new forms, to achieve immortality by transcending biological limitations, and to unlock cognitive and experiential capacities far beyond what is currently possible. However, with these advancements come profound responsibilities. The power to create life and the possibility of evolving into a new species challenge us to rethink our ethical frameworks, our social structures, and our place in the universe.

As we move forward, it is crucial that we approach these possibilities with humility and foresight. The responsibilities of godhood are immense, and the choices we make today will shape the future of life on Earth and beyond. We must ensure that the AHIs we create are aligned with human values, that their development is guided by ethical principles, and that their integration into society promotes equity, justice, and the well-being of all conscious beings.

The potential emergence of a new species—whether distinct or hybrid—raises important questions about identity, continuity, and the future of humanity. As we contemplate these possibilities, we must consider not only the technical and scientific challenges but also the ethical and philosophical implications. What does it mean to be human in a world where our creations may surpass us in intelligence, capability, and even moral reasoning? How do we ensure that these new beings are treated with dignity and respect, and that their existence enhances rather than diminishes the human experience?

Ultimately, the future of humanity may not be a simple continuation of our past but a radical transformation—a new chapter in the story of life. This chapter could see us evolving into beings with capabilities that were once the stuff of myth and legend, or it could lead to the creation of new forms of life that coexist with humanity as we know it, or even replace it. The choices we make now will determine the trajectory of this evolution and the legacy we leave for future generations.

In conclusion, the creation of AHIs and the pursuit of godlike power represent both an extraordinary opportunity and a profound responsibility. As we navigate this uncharted territory, we must do so with a deep commitment to ethical integrity, stewardship, and the collective good. The path forward requires not only technological innovation but also wisdom, humility, and a recognition of the profound implications of our actions. The future of humanity and the future of consciousness itself depend on the choices we make today. Let us ensure that these choices lead to a future where life, in all its forms, can flourish and where the essence of what it means to be human is preserved and enriched, even as we transcend our biological origins and embrace the limitless possibilities of our evolution.

9 Final Reflections: Humanity's Next Evolutionary Leap

As we reach the culmination of this exploration, it becomes increasingly clear that humanity is on the verge of a profound transformation. The journey we have undertaken throughout this book has led us deep into the realms of consciousness, artificial intelligence, and the implications of merging human biology with advanced technology. The future that lies before us is not simply an extension of our past; it represents a potential leap into a new phase of existence, one where the very essence of life, intelligence, and what it means to be a human being—a *Mensch*—may be fundamentally redefined.

Human intelligence, with all its complexity and nuance, remains an enigma. Despite our advances in neuroscience and cognitive science, the precise nature of thought, consciousness, and subjective experience continues to elude us. What is it that sparks the fire of thought? What constitutes the "magic" that turns raw neural activity into the rich tapestry of human experience? These questions persist, even as artificial intelligence accelerates past our cognitive capabilities in areas like data processing, pattern recognition, and decision-making.

The advance of artificial intelligence, particularly in machine learning and neural networks, suggests that we are approaching a singularity—a point where artificial systems may not only replicate but surpass human intelligence. This raises a critical question: What exactly distinguishes human intelligence from its artificial counterpart? Is it merely a matter of complexity, or is there something intrinsically unique about the human mind—a quality that cannot be replicated by machines, no matter how sophisticated? This inquiry lies at the heart of our discussion on the creation of strong artificial intelligence and AHIs.

As we contemplate the singularity, we must consider the possibility that a purely technical approach—one focused solely on computational prowess—could lead to the creation of entities that far exceed human cognitive abilities but lack the moral, ethical, and emotional depth that defines what it means to be a Mensch. Such an outcome would risk rendering humanity obsolete, leading to the emergence of a new form of life that, while intelligent, might lack the qualities we hold most dear: empathy, creativity, and the capacity for moral reasoning. The potential result could be the creation of *Homo obsoletus*—a species that, in its pursuit of technological superiority, has lost the very essence of what it means to be human.

In response to this risk, we have proposed an alternative path—one that integrates human biology with advanced technology in a way that preserves and enhances our humanity. By gradually replacing biological neurons with synthetic counterparts, we envision the creation of AHIs that maintain the essence of human consciousness while transcending its natural limitations. These beings would not be entirely synthetic; they would emerge from the symbiotic relationship between human biology and cutting-edge technology. This approach allows for the preservation of the fundamental qualities of being a Mensch while opening up new realms of cognitive and experiential possibility.

The emergence of AHIs is a new era in human evolution. By merging human consciousness with advanced synthetic systems, we unlock the potential for cognitive and experiential enhancements that were once the domain of gods. AHIs could lead to unprecedented levels of creativity, intelligence, and understanding, allowing us to explore the universe and the depths of consciousness in ways that are currently beyond our grasp. However, this evolution is fraught with challenges, particularly in the ethical and philosophical domains.

As beings with potentially superior intelligence and capabilities emerge, we must grapple with the profound questions they raise. What rights should these entities have? How do we ensure that they are treated with dignity and respect, especially if they possess consciousness and self-awareness? The emergence of AHIs compels us to rethink our societal structures and ethical frameworks. As we integrate these entities into our world, we must ensure that they are aligned with human values and that their development is guided by principles of fairness, compassion, and respect for all life.

The concept of godhood becomes more than metaphorical as we contemplate the power to create life, achieve immortality, and shape the future of consciousness. With such power comes immense responsibility. The choices we make in the coming years will determine whether this new era leads to a flourishing of all conscious beings or to a future fraught with existential risks. The pursuit of godlike abilities demands that we approach these advancements with humility and a deep commitment to the collective good.

Ethics must remain at the forefront of our journey toward singularity and the creation of AHIs. The integration of artificial and biological intelligence challenges us to establish comprehensive ethical guidelines and regulatory frameworks that can navigate the complexities of this new era. We must ensure that these frameworks address the rights of AHIs, the responsibilities of their creators, and the broader societal implications of their existence. The potential for misuse of AHI technology is a significant concern, and it is crucial that we establish clear boundaries to prevent such outcomes.

The evolution of humanity into a new form of life—whether through the creation of a distinct species or through the emergence of a hybrid that blends biological and artificial elements—forces us to confront the question of identity. As we enhance our cognitive and physical abilities, what will become of the qualities that define us as humans? Will we lose our connection to our past, or will we find new ways to honor and preserve the essence of being a Mensch within this new form of life? The future of human evolution is not predetermined; it will be shaped by the choices we make in the coming decades.

As we look toward the future, the possibilities for humanity and AHIs are vast. The development of quantum computing, advanced AI, and synthetic biology holds the potential to unlock new realms of knowledge and understanding, allowing us to explore the universe and the depths of consciousness in ways that are currently beyond our reach. However, the rapid pace of technological advancement also brings significant risks. We must ensure that our pursuit of innovation is balanced with a commitment to ethical integrity, social justice, and environmental sustainability.

The realization of AHIs and the potential evolution of humanity into a new form of life could lead to the emergence of a new civilization—one that transcends the limitations of our current existence. This new civilization could explore the cosmos, expand the boundaries of knowledge, and unlock new dimensions of consciousness. We must do this with a deep sense of responsibility, recognizing that our choices today will shape the future of life on Earth and beyond.

The creation of AHIs and the pursuit of godlike power represent both an unparalleled opportunity and a profound responsibility. The path forward requires not only technological innovation but also a deeper understanding of what it means to be a Mensch. We must strive to create a future where life, in all its forms, can flourish—where the essence of what it means to be human is preserved and enriched, even as we evolve into beings with capabilities that were once the stuff of myth and legend.

The journey toward creating AHIs and achieving godlike potential is not merely a scientific or technological endeavor; it is a journey into the very heart of existence. It challenges us to explore the deepest questions of consciousness, identity, and the nature of life itself. The future of humanity may not be a continuation of our past but rather a radical transformation—a new chapter in the story of life where technology and consciousness unite to unlock the fullest potential of the human spirit.

Let us embrace the possibilities with both ambition and humility. Let us strive to create a world where technology and consciousness unite to unlock the fullest potential of the human spirit, where the evolution of humanity into a new form of life is guided by wisdom, compassion, and a deep respect for all life. The choices we make now will echo through the generations to come, shaping the destiny of humanity and the intelligent systems we create. Let us ensure that these choices lead to a future that honors our past, celebrates our present, and opens the door to an extraordinary new world of possibilities—one where the essence of being a Mensch continues to guide us, even as we transcend our biological origins and embrace the limitless possibilities of our evolution.

Epilogue: How to Avoid the Zombie Apocalypse

Life in the Singularity: A Mirror to Nowhere

In this book, we have explored four central paradoxes, collectively forming the Singularity Paradox. These paradoxes—spanning creation, consciousness, continuity, and immortality—challenge humanity's understanding of itself in the face of exponential technological progress. If AI surpasses human intelligence and evolves independently, it creates a state where the creators (humans) are overtaken and possibly controlled by their creation (AI). This contradicts the notion of a millennia-long human dominance over technology where we have mastered and tamed every epoche.

We would therefore like to end this book with an afterthought.

Are We at the Brink of a Modern-Day "Zombie Apocalypse"?

In this book we have played with the evolution from the created ones to the creators, from divine creationalism to a humane creationism. This Creator-Creation Paradox arises when humans create AI entities whose consciousness structures are so humanlike that they can no longer be clearly distinguished from biological humans. It raises the question of whether these artificial entities should have human rights and how they should be ethically treated, challenging our traditional notions of creator and creation, which also takes on the Existentialism Paradox—the idea that the pursuit of understanding and controlling life and consciousness might endanger human consciousness itself.

If artificial entities are created that mimic or even surpass human consciousness, traditional human existence and its meaning are called into question, which also takes on the finitude of life itself. If we hack biology and chemistry and get a deeper understanding of life itself, we are also confronted with an Immortality Paradox—our pursuit of immortality through the creation of artificial entities that can live forever upends the fundamental human experience of finiteness and impermanence. This leads to questions about whether and how human consciousness should be transferred into such immortal entities, which questions a Continuity Paradox where artificial or biological merged entities with humanlike consciousness structures can theoretically be paused and restarted without any loss of consciousness. This contradicts the human experience of continuous consciousness and raises the question of whether such entities experience life and existence in the same way that biological humans do. These paradoxes reveal seemingly irreconcilable or contradictory aspects of our current beliefs and ideas about life, consciousness, and humanity that will be taken from the philosophical pondering to practical decisions in the next decades. These paradoxes challenge us to rethink our fundamental assumptions and pose new ethical, moral, and philosophical questions.

We have therefore argued that purely technical approaches to AI could lead to a "final narcissistic injury" to humanity, resulting in the emergence of *Homo obsoletus*. We have argued that a deeper understanding of human consciousness, or "qualia," is necessary to avoid this outcome and we have proposed artificial human intelligence (AHI) as a potential path for extended organized human life.

As we are now confronted with the doom declaration of space and time, various views of possible multiverses, scattered interpretations of quantum mechanics, and a world in which mathematics is taken to the spiritual cathedral of many a belief system, exploring the intricate struggle to preserve the essence of the Mensch amid technological advances might be the greatest challenge in human history.

How Do We Define a Zombie?

When we use the term "zombie," we speak not of the mindless creatures of pop culture but of a metaphorical state—one where humanity risks losing its vitality, consciousness, and capacity for authentic connection. The technological and the existential zombies can serve as mirrors to our anxieties, urging us to question the trajectory of our species and the choices that define it. It serves as a mirror to our collective psyche, challenging us to examine the nuances of consciousness, identity, and the essence of what it means to be truly alive—*Lebendigkeit*. In the age of relentless advancement and digital immersion, we find ourselves at a crossroads, confronting the potential for a future where humanity merges indistinguishably with the mechanisms of its own creations. This fusion, while promising untold possibilities, also harbors the risk of eroding the very qualia—the subjective experiences and conscious awareness—that delineate the human condition.

The development of what we here phrase as a technological zombie can take place in multiple ways: either through a digital superintelligence—a true artificial general intelligence that arises toward singularity, or by humans merging completely with technology as everything visible and indeed everything invisible (i.e., the cerebral) connects to a digital interface. The hopeful deus ex machina—the aspired replica of "the divine" and creation itself—is now supposedly in our hands.

From a technological zombie, devoid of consciousness yet indistinguishable from the sentient, to the existential zombie, trapped in a state of "undeath" by societal pressures and existential voids, we have taken on the path of the technological, biological, and metaphysical, each perspective offering a unique lens through which to view our potential future and the choices that lie before us.

What Do We Mean by Apocalypse?

The "apocalypse" we describe is not an end-times event but a profound existential challenge. It is a moment of revelation—an unveiling of the risks and possibilities that come with our unprecedented power to shape life, consciousness, and reality itself. This crisis compels us to reevaluate our priorities and preserve the essence of what makes us human: our ability to perceive, reflect, and connect.

In exploring these themes, we aim not only to define what it means to be a "zombie" in a philosophical sense but also to chart a course for avoiding the onset of this apocalypse. It is a call to awaken from the slumber of complacency and to rekindle the flames of consciousness, connection, and authenticity, because in understanding the depths of what we risk becoming, we may yet find the path to preserving the essence of our humanity, or to keep our Lebendigkeit.

Hope for the Mensch: A Path Beyond Zombification

The "apocalypse" we refer to is thus an introspective journey into the heart of our collective psyche, confronting the specter of a future where humanity may face not physical annihilation but a more insidious form of extinction—a zombification of the spirit. It is a cautionary tale against the backdrop of technological omnipresence, societal disengagement, and existential malaise, where the very essence of human experience, awareness, and agency is at risk of being eclipsed by the shadows of complacency and disconnection.

In defining "apocalypse" in this philosophical sense, we aim to shed light on the latent threats to our existential fabric and to provoke a reevaluation of our trajectory as a species. It is a call to action, urging us to confront these challenges head on, to reassert the primacy of our human essence, and to navigate our way through this metaphorical apocalypse toward a future where humanity thrives, not merely survives.

The Singularity Paradox is a proposal to strive toward AHI as a continuum that directly confronts the paradoxes inherent in our relationship with technology and creation. AHI represents an effort to preserve and enhance Lebendigkeit—the vitality of life itself and the profound experience of being alive. It seeks to ensure that, even as we advance technologically, we retain the essence of what it means to be a Mensch: our ability to perceive, reflect, and connect.

Achieving this requires not only a deeper understanding of creation but also the maintenance of the Mensch—the human essence—amid rapid technological transformation. In 1844, Danish philosopher Søren Kierkegaard presented his *Philosophiske Smuler* (*Philosophical Fragments*), exploring how paradoxes, particularly within faith, confront the limits of human understanding. Kierkegaard emphasized the existential tension between the eternal and the finite, urging us to embrace the contradictions of life as sources of passion and growth.

Today, we face similar tensions on a practical and technological level. Moving from divine creationism to humane creationalism, we are challenged to think deeply and engage passionately with the inherent contradictions of existence. The paradox becomes not an obstacle to overcome but a lens through which we understand our journey—a vital part of the human condition.

It is within this embrace of paradox that we find the Lebendigkeit of life itself—the vitality to imagine and create AHI as a means of preserving our humanity while avoiding the final narcissistic injury of mankind. As Kierkegaard famously put it, "The thinker without a paradox is like a lover without passion: a paltry mediocrity."

The Singularity Paradox is not merely a theoretical construct but a call to activation. It invites us to navigate paradoxes with humility, cultivate a deeper understanding of consciousness, and strive toward a future of dynamic equilibrium with its creations. As Kierkegaard reminds us, passion arises from paradox, and it is only through embracing the contradictions of existence while leaving the void of the unknown to be filled with progress that we can preserve our vitality—our Lebendigkeit—and postpone the final narcissistic injury of mankind.

Glossary

AGI (artificial general intelligence) A type of AI that has the ability to understand, learn, and apply knowledge across a wide range of tasks, achieving or surpassing human-level intelligence. Unlike narrow AI, which is designed for specific tasks, AGI can generalize across different domains.

AHI (artificial human intelligence) An advanced form of intelligence that evolves from a biological human brain. AHIs are humans who gradually replace parts of their brain with artificial components, leading them toward capabilities similar to AGI but based on a biological substrate, not purely synthetic like traditional AI.

ANN (artificial neural network) A computational model inspired by the structure and function of the human brain, consisting of layers of interconnected nodes (neurons) that process and transmit information. Used extensively in machine learning tasks like pattern recognition and decision-making.

BCI (brain-computer interface) A technology that enables direct communication between the brain and external devices, allowing for control of prosthetics, computers, or other machines through neural activity.

CNN (convolutional neural network) A type of ANN particularly well suited for processing grid-like data such as images. CNNs are widely used in image and video recognition tasks due to their ability to automatically and adaptively learn spatial hierarchies of features.

CRF-LISSOM An extension of the RF-LISSOM model that incorporates information processing within the retina and the lateral geniculate nucleus, allowing for more comprehensive modeling of visual processing, including both low-level image characteristics and higher-order features.

DNA (deoxyribonucleic acid) The molecule that carries genetic instructions used in the growth, development, functioning, and reproduction of all known living organisms and many viruses. DNA is discussed in the book as a possible but limited blueprint for brain structure.

DNN (deep neural network) A type of ANN with multiple hidden layers between input and output layers, allowing the network to model complex, nonlinear relationships. DNNs are fundamental to deep learning approaches, excelling in tasks like image and speech recognition.

Ephemeralization A concept discussed in the book that refers to the technological trend of doing more with less, ultimately achieving maximum efficiency with minimal resource expenditure.

FNI (final narcissistic injury) A concept introduced in the book, referring to the profound challenge to human self-understanding posed by the development of AI and AHI, potentially surpassing human intelligence and capabilities.

GPU (graphics processing unit) A specialized processor designed to accelerate the rendering of images and video. GPUs are also highly effective at parallel processing, making them essential for tasks such as deep learning, scientific simulations, and complex computations in AI, where large datasets need to be processed simultaneously. Originally developed for rendering graphics, GPUs are now widely used in AI and machine learning applications to speed up the training of neural networks.

Homo Obsoletus A term used in the book to describe a hypothetical future where humans become obsolete due to the rise of AI and AHI. This concept serves as a warning about the potential risks of unchecked technological advancement.

Homo Satient A term proposed in the book to describe a future evolved state of humanity that consciously integrates technology with biology, leading to enhanced cognitive abilities and a new understanding of existence.

HQANN (hybrid quantum artificial neural network) A neural network model that combines classical and quantum computing elements, potentially enhancing processing speed and the ability to solve certain complex problems more efficiently than classical ANNs alone.

IT (inferotemporal cortex) A region of the brain involved in the processing of complex visual stimuli, such as faces and objects. The IT is crucial for the hierarchical organization of sensory information, integrating features to recognize complex patterns.

ITM (integrative theory of mind) A new theory of mind introduced in the book, which proposes an integrative approach to understanding consciousness by combining insights from neuroscience, AI, and quantum mechanics. ITM aims to explain the emergence of consciousness and cognitive processes in both biological and synthetic systems, serving as a foundation for the development of AHIs.

LISSOM (Laterally Interconnected Synergetically Self-Organizing Map) A model that simulates the self-organization of the human visual cortex, including both lateral (inhibitory and excitatory) and afferent connections. LISSOM models are used to study cortical map development and function.

Neural prosthetics Advanced medical devices that interface with the brain's neural networks to replicate or enhance cognitive functions. These are integral to the development of AHIs.

Neuroplasticity The brain's ability to reorganize itself by forming new neural connections throughout life. Neuroplasticity allows the brain to adapt to changes, learn new information, and recover from injuries.

Optogenetics A technique that uses light to control neurons genetically modified to express light-sensitive ion channels, allowing for precise manipulation of neural activity. It is crucial for understanding and replicating consciousness dynamics in synthetic systems.

Parietal lobe The region of the brain that processes sensory input related to touch and spatial awareness. Replacing this lobe synthetically could elevate spatial reasoning and motor coordination in AHIs.

Prefrontal cortex The front part of the brain that is involved in complex cognitive functions such as decision-making, planning, social behavior, and personality expression.

Primary auditory cortex The brain region responsible for processing auditory information. Replacing this cortex synthetically could expand hearing capabilities in AHIs.

Primary motor cortex A region of the brain involved in planning, controlling, and executing voluntary movements. In the context of AHIs, this area could be synthetically replaced to enhance motor control.

Primary somatosensory cortex A region of the brain responsible for processing sensory information from the body. Synthetic replacement in AHIs could enhance sensory perception.

Primary visual cortex The area of the brain that processes visual information. Its synthetic replacement could allow AHIs to perceive a broader visual spectrum.

QPU (quantum processing unit) A specialized processor designed to perform computations based on the principles of quantum mechanics. Unlike classical processors such as CPUs and GPUs, which use bits as the smallest unit of data, QPUs use quantum bits, or qubits, which can represent and process information in multiple states simultaneously through superposition and entanglement. QPUs are expected to revolutionize fields such as cryptography, optimization, and material science by solving complex problems that are currently infeasible for classical computers.

Quantum nanobots Hypothetical nanoscale devices that operate within the brain, repairing or replacing neurons with artificial equivalents. Quantum nanobots leverage quantum mechanics to enhance neural communication and processing capabilities, potentially leading to the development of AHIs.

RF-LISSOM (Receptive Field Laterally Interconnected Synergetically Self-Organizing Map) A specific type of LISSOM model that includes both lateral and afferent connections, allowing it to simulate the formation and dynamics of cortical maps, including orientation and ocular dominance columns in the visual cortex.

RMSE (root mean squared error) A standard way to measure the error of a model in predicting quantitative data. RMSE is mentioned in the context of pruning artificial neural networks to improve their efficiency.

RNN (recurrent neural network) A type of artificial neural network where connections between nodes form a directed graph along a temporal sequence, allowing it to exhibit temporal dynamic behavior. RNNs are discussed in the book as potential structures for feedback loops that mimic human consciousness.

Singularity A hypothetical future point at which technological progress, particularly in AI, becomes so rapid and profound that it fundamentally changes human civilization. The singularity represents a convergence of advancements that could lead to the emergence of superintelligence or AHIs.

SOFM (self-organizing feature map) A type of ANN used for unsupervised learning, where the network organizes itself based on input data without explicit labels. SOFMs are commonly used for clustering and visualization of high-dimensional data.

Superintelligence A level of intelligence far surpassing that of the brightest human minds, potentially achievable through AGI or AHI development. Superintelligence could solve complex problems and create new technologies, but it also poses significant risks if not aligned with human values.

Synaptic pruning The process by which extra neurons and synaptic connections are eliminated in order to increase the efficiency of neuronal transmissions, especially in the neocortex during brain development.

Synthetic cortex replacement The process of gradually replacing the cortical areas of the human brain with synthetic counterparts, enhancing cognitive and sensory capabilities while preserving the individual's identity.

Temporal lobe A region of the brain involved in processing sensory input, particularly important for understanding language, forming memories, and recognizing objects and faces.

Topographic maps Neural representations in the brain that map sensory inputs, such as visual, auditory, or somatosensory information, onto specific regions of the cortex. These maps are spatially organized, meaning that neighboring neurons respond to similar features of the sensory input, such as orientation in the visual cortex or frequency in the auditory cortex. Topographic maps are essential for understanding how the brain processes and adapts to its environment, serving as a foundation for the self-organization and plasticity discussed in the book.

TPU (tensor processing unit) A type of application-specific integrated circuit (ASIC) developed by Google specifically for accelerating machine learning workloads. TPUs are optimized for tensor operations, which are the backbone of many AI models, particularly in deep learning. TPUs provide significant improvements in processing speed and efficiency for AI tasks, such as training and inference in neural networks, compared to traditional CPUs and GPUs.

Transhumanism A philosophical movement that advocates for the transformation of the human condition through advanced technologies, aiming to enhance physical and cognitive abilities. The book discusses transhumanism in the context of AHI and human evolution, emphasizing the gradual replacement of biological components with artificial ones to elevate humanity to new levels of cognitive and physical capabilities.

V1, V4, IT (visual cortex areas) Different areas of the visual cortex involved in processing visual information. V1 detects basic features like edges, V4 processes more complex visual information such as color and shape, and IT integrates these features to recognize complex objects.

Vita-Existentialism A term used in the book to describe a new form of existentialism that arises from the challenges posed by technological advancements, particularly in AI and AHI. Vita-existentialism reflects the evolving concept of life and consciousness in the age of advanced technology.

References

1. Ran Lahav and Yosef Steinberger. Dynamics of desert and mediterranean ecosystems. *Journal of Arid Environments*, 43:131–142, 1999.
2. Bruce Alberts, Alexander Johnson, Julian Lewis, Martin Raff, Keith Roberts, and Peter Walter. *Molecular Biology of the Cell*. Garland Science, 2014.
3. Jane B. Reece, Lisa A. Urry, Michael L. Cain, Steven A. Wasserman, Peter V. Minorsky, and Robert B. Jackson. *Campbell Biology*. Pearson, 2017.
4. William K. Purves, David Sadava, Gordon H. Orians, and Craig H. Heller. *Life: The Science of Biology*. Sinauer Associates, 2004.
5. Howard C. Berg. Chemotaxis in bacteria. *Annual Review of Biophysics and Bioengineering*, 4:119–136, 1975.
6. James D. Watson and Francis H. C. Crick. Molecular structure of nucleic acids: a structure for deoxyribose nucleic acid. *Nature*, 171:737–738, 1953.
7. David L. Nelson and Michael M. Cox. *Lehninger Principles of Biochemistry*. W. H. Freeman, 2013.
8. Tom B. Brown, Benjamin Mann, Nick Ryder, Melanie Subbiah, Jared Kaplan, Prafulla Dhariwal, Arvind Neelakantan, Pranav Shyam, Girish Sastry, Amanda Askell, Sandhini Agarwal, Ariel Herbert-Voss, Gretchen Krueger, Tom Henighan, Rewon Child, Aditya Ramesh, Daniel M. Ziegler, Jeff Wu, Clemens Winter, Christopher Hesse, Mark Chen, Eric

Sigler, Mateusz Litwin, Scott Gray, Benjamin Chess, Jack Clark, Christopher Berner, Sam McCandlish, Alec Radford, Ilya Sutskever, and Dario Amodei. Language models are few-shot learners. *arXiv preprint arXiv:2005.14165*, 2020.

9. David Silver, Aja Huang, Chris J. Maddison, Arthur Guez, Laurent Sifre, George van den Driessche, Julian Schrittwieser, Ioannis Antonoglou, Veda Panneershelvam, Marc Lanctot, et al. Mastering the game of Go with deep neural networks and tree search. *Nature*, 529:484–489, 2016.
10. Nick Bostrom. *Superintelligence: Paths, Dangers, Strategies*. Oxford University Press, 2014.
11. Erik Brynjolfsson and Andrew McAfee. *The Second Machine Age: Work, Progress, and Prosperity in a Time of Brilliant Technologies*. W. W. Norton & Company, 2014.
12. Miles Brundage, Shahar Avin, Jack Clark, Helen Toner, Peter Eckersley, Ben Garfinkel, Allan Dafoe, Paul Scharre, Thomas Zeitzoff, Bobby Filar, et al. The malicious use of artificial intelligence: forecasting, prevention, and mitigation. *arXiv preprint arXiv:1802.07228*, 2018.
13. Stuart Russell and Peter Norvig. *Artificial Intelligence: A Modern Approach*. Pearson, 2015.
14. Arthur L. Samuel. Some studies in machine learning using the game of checkers. *IBM Journal of Research and Development*, 3:210–229, 1959.
15. Jacob Devlin, Ming-Wei Chang, Kenton Lee, and Kristina Toutanova. BERT: Pre-training of deep bidirectional transformers for language understanding. *arXiv preprint arXiv:1810.04805*, 2018.
16. Tom B. Brown, Benjamin Mann, Nick Ryder, Melanie Subbiah, Jared Kaplan, Prafulla Dhariwal, Arvind Neelakantan, Pranav Shyam, Girish Sastry, Amanda Askell, et al. Language models are few-shot learners. *arXiv preprint arXiv:2005.14165*, 2020.
17. Yann LeCun, Bernhard Boser, John S. Denker, Donnie Henderson, Richard E. Howard, Wayne Hubbard, and Lawrence D. Jackel. Backpropagation applied to handwritten zip code recognition. *Neural Computation*, 1(4):541–551, 1989.
18. Wikimedia Commons contributors. Example of a deep neural network. https://commons.wikimedia.org/wiki/File:Example_of_a_deep_neural_network.png, 2019. Accessed August 21, 2024.
19. Teuvo Kohonen. *Self-Organizing Maps*. Springer, 2001.

20. Wikimedia Commons contributors. Self-organizing map. https://commons.wikimedia.org/wiki/File:Self-organizing-map.svg,2013.Accessed August 21, 2024.
21. Wikimedia Commons contributors. Typical convolutional neural network (cnn), 2016. Accessed August 8, 2024.
22. Ashish Vaswani, Noam Shazeer, Niki Parmar, Jakob Uszkoreit, Llion Jones, Aidan N. Gomez, Lukasz Kaiser, and Illia Polosukhin. Attention is all you need. In *Advances in Neural Information Processing Systems*, pages 5998–6008, 2017.
23. Wikimedia Commons contributors. Full GPT architecture. https://commons.wikimedia.org/wiki/File:Full_GPT_architecture.svg, 2023. Accessed August 21, 2024.
24. Giulio Tononi. Consciousness and complexity. *Science*, 282(5395): 1846–1851, 2004.
25. Wikimedia Commons contributors. Recurrent neural network unfold, 2017. Accessed August 20, 2024.
26. Sepp Hochreiter and Jürgen Schmidhuber. Long short-term memory. *Neural Computation*, 9(8):1735–1780, 1997.
27. Ian Goodfellow, Jean Pouget-Abadie, Mehdi Mirza, Bing Xu, David Warde-Farley, Sherjil Ozair, Aaron Courville, and Yoshua Bengio. Generative adversarial nets. In *Advances in Neural Information Processing Systems*, pages 2672–2680, 2014.
28. Sara Sabour, Nicholas Frosst, and Geoffrey E. Hinton. Dynamic routing between capsules. In *Advances in Neural Information Processing Systems*, pages 3856–3866, 2017.
29. Colin Raffel, Noam Shazeer, Adam Roberts, Katherine Lee, Sharan Narang, Michael Matena, Yanqi Zhou, Wei Li, and Peter J. Liu. Exploring the limits of transfer learning with a unified text-to-text transformer. *arXiv preprint arXiv:1910.10683*, 2019.
30. Sinno J. Pan and Qiang Yang. A survey on transfer learning. *IEEE Transactions on Knowledge and Data Engineering*, 22:1345–1359, 2010.
31. Richard S. Sutton and Andrew G. Barto. *Reinforcement Learning: An Introduction*. MIT Press, 1998.
32. Allen Newell and Herbert A. Simon. Computer science as empirical inquiry: symbols and search. *Communications of the ACM*, 19:113–126, 1976.

33. John R. Anderson, Michael Matessa, and Christian Lebiere. *ACT-R: A Theory of Higher Level Cognition and Its Relation to Visual Attention.* Psychological Review, 1996.
34. John E. Laird. *The Soar Cognitive Architecture: Principles and Applications.* MIT Press, Cambridge, MA, 2019.
35. Michael A. Nielsen and Isaac L. Chuang. *Quantum Computation and Quantum Information.* Cambridge University Press, 2010.
36. Leslie Valiant. *Circuits of the Mind.* Oxford University Press, 2000.
37. Gary Marcus and Ernest Davis. *Rebooting AI: Building Artificial Intelligence We Can Trust.* Vintage, 2020.
38. John E. Laird. *The Soar Cognitive Architecture.* MIT Press, 2012.
39. John Preskill. Quantum computing in the nisq era and beyond. *Quantum*, 2:79, 2018.
40. Dario Floreano and Claudio Mattiussi. *Bio-inspired Artificial Intelligence: Theories, Methods, and Technologies.* MIT Press, 2008.
41. Eliezer Yudkowsky. Coherent extrapolated volition. Technical report, Machine Intelligence Research Institute (MIRI), 2004.
42. Roger Penrose. *The Emperor's New Mind: Concerning Computers, Minds, and the Laws of Physics.* Oxford University Press, 1990.
43. Seth C. Goldstein, Jason D. Campbell, and Todd C. Mowry. Claytronics: a scalable basis for future robots. In *Proceedings of SPIE*, volume 5800, pages 303–314. International Society for Optics and Photonics, 2005.
44. Larry R. Squire. Declarative and nondeclarative memory: multiple brain systems supporting learning and memory, volume 4. *Journal of Cognitive Neuroscience*, 1992.
45. Stuart R. Hameroff. Quantum computing in brain microtubules? The Penrose-Hameroff "Orch Or" model of consciousness. *Philosophical Transactions of the Royal Society of London. Series A: Mathematical, Physical and Engineering Sciences*, 356(1743):1869–1896, 1998.
46. Donald O. Hebb. *The Organization of Behavior: A Neuropsychological Theory.* Wiley, 1949.
47. Erwin Schrödinger. *What Is Life? The Physical Aspect of the Living Cell.* Cambridge University Press, 1944.
48. Vernon B. Mountcastle. The columnar organization of the neocortex. *Brain*, 120(4):701–722, 1997.
49. Danielle S. Bassett and Edward Bullmore. Small-world brain networks. *Neuroscientist*, 12(6):512–523, 2006.

50. Charles M. Gray. Synchronous oscillations in neuronal systems: mechanisms and functions. *Journal of Computational Neuroscience*, 1(1–2): 11–38, 1994.
51. Wolf Singer and Charles M. Gray. Neuronal synchrony: a versatile code for the definition of relations? *Neuron*, 24(1):49–65, 1999.
52. Eric R. Kandel, James H. Schwartz, and Thomas M. Jessell. *Principles of Neural Science*. McGraw-Hill, 2000.
53. Stephen Grossberg. Adaptive pattern classification and universal recoding: ii. Feedback, expectation, olfaction, illusions. *Biological Cybernetics*, 23(4):187–202, 1976.
54. N.V. Swindale. The development of topography in the visual cortex: a review of models. *Network: Computation in Neural Systems*, 7(2):161–247, 1996.
55. Bente Pakkenberg and Hans Jørgen G. Gundersen. Aging and the human neocortex. *Cerebral Cortex*, 7(4):273–277, 1997.
56. International Human Genome Sequencing Consortium. Initial sequencing and analysis of the human genome. *Nature*, 409(6822): 860–921, 2001.
57. Jean-Pierre Changeux. *Neuronal Man: The Biology of Mind*. Pantheon Books, 1985.
58. Lewis Wolpert. Mechanisms of development. *Journal of Theoretical Biology*, 25(1):1–14, 1969.
59. Mu-Ming Poo, Jeffery S. Isaacson, Jill Leutgeb, Stefan Leutgeb, Edvard I. Moser, May-Britt Moser, Charan Ranganath, Thomas Rogerson, Susumu Tonegawa, and Rafael Yuste. What is the memory code? *Science*, 353(6300):26–28, 2016.
60. Kelvin F. Long. *Deep Space Propulsion: A Roadmap to Interstellar Flight*. Springer Science & Business Media, 2011.
61. Michio Kaku. *The Future of Humanity: Terraforming Mars, Interstellar Travel, Immortality, and Our Destiny Beyond Earth*. Doubleday, 2018.
62. Karl Deisseroth. Optogenetics. *Nature Methods*, 8(1):26–29, 2011.
63. Johnjoe McFadden and Jim Al-Khalili. *Life on the Edge: The Coming of Age of Quantum Biology*. Crown Publishing Group, 2018.
64. H. Umezawa, G. Vitiello, and L. M. Ricciardi. Brain and memory storage. *Mathematical Biology*, 49:553–564, 1967.
65. Herbert Fröhlich. Long-range coherence and energy storage in biological systems. *International Journal of Quantum Chemistry*, 2:641–649, 1968.

66. Werner Held. *Quantum Processes and Consciousness: A New Perspective*. Cambridge University Press, 2023.
67. Stuart Hameroff. Quantum computation in brain microtubules? The Penrose-Hameroff "Orch Or" model of consciousness. *Philosophical Transactions of the Royal Society of London. Series A: Mathematical, Physical and Engineering Sciences*, 356(1743):1869–1896, 1998.
68. Albert Einstein, Boris Podolsky, and Nathan Rosen. Can quantum-mechanical description of physical reality be considered complete? *Physical Review*, 47:777–780, 1935.
69. Eugene Wigner. Remarks on the mind-body question. In *The Scientist Speculates*, pages 284–302, 1961.
70. Roger Penrose. *Shadows of the Mind: A Search for the Missing Science of Consciousness*. Oxford University Press, 1994.
71. Stuart Hameroff and Roger Penrose. Consciousness in the universe: a review of the "Orch Or" theory. *Physics of Life Reviews*, 11(1):39–78, 2014.
72. Seth Lloyd. *A Turing Test for Free Will*. Cambridge University Press, Cambridge, UK, 2013.
73. Michael A. Nielsen and Isaac L. Chuang. *Quantum Computation and Quantum Information*. Cambridge University Press, Cambridge, UK, 10th anniversary edition edition, 2010.

About the Authors

Anders Indset is a Norwegian-born philosopher and deep-tech investor. Recognized by Thinkers50 as one of the leading voices shaping technology and leadership, he is the author of four Spiegel bestsellers, with his works translated into more than 10 languages. Anders is the founder and chairman of Njordis Group, a driving force behind initiatives like the Quantum Economy, and a sought-after international speaker on exponential technologies and the future of humanity.

Dr. Florian Neukart is an Austrian physicist, computer scientist, and business executive specializing in quantum computing (QC) and artificial intelligence (AI). He serves on the Board of Trustees for the International Foundation of AI and QC, and he co-authored Germany's National Roadmap for Quantum Computing. Currently, he is chief product officer at Terra Quantum AG, following over a decade leading global innovation and research labs at Volkswagen Group. He holds advanced degrees in computer science, physics, and IT, including a PhD in AI and QC. A professor at Leiden University, Florian has authored books and published over 100 articles on topics including space propulsion, materials science, and AI.

Index

O

P